T0329687

INTRODUCTION TO WIRELESS SENSOR NETWORKS

INTRODUCTION TO WIRELESS SENSOR NETWORKS

ANNA FÖRSTER

IEEE PRESS

WILEY

For general information on our other products and services or for technical support, please contact our Customer Care Department within the United States at (800) 762-2974, outside the United States at (317) 572-3993 or fax (317) 572-4002.

Wiley also publishes its books in a variety of electronic formats. Some content that appears in print may not be available in electronic formats. For more information about Wiley products, visit our website at www.wiley.com.

Library of Congress Cataloging-in-Publication Data is available.

ISBN: 978-1-118-99351-4

Printed in the UK

To my parents, Radmila and Alexey, who ignited my love for science and computers.

CONTENTS

HOW TO USE THIS BOOK

Let us begin by exploring how to use this book and what tools and prior knowledge you will need to learn about wireless sensor networks. The following sections discuss the tools, software, and hardware that are needed as well as where to find them. Additionally, your needed level of programming experience is discussed. If you do not yet have programming experience, resources to obtain it are also offered. Some of the resources are listed here in this book but many more are also available online at the official book website:

comnets.uni-bremen.de/wsn-book/

WHO IS THIS BOOK FOR?

This book assumes that you have more interest in wireless sensor networks (WSNs) than simply buying a ready solution and installing it. Thus, the book is perfect for wireless sensor networks beginners interested in how WSNs work, how to implement them, and how to do research in WSNs. It is also well suited for students both at the undergraduate and graduate levels, programming experts entering the topic of WSNs, as well as hobbyists interested in building their own WSNs.

HOW TO READ THIS BOOK?

Reading this book from the beginning to the end is the best choice. Each chapter starts with an overview of what you will learn and ends with a Summary and

Further Reading section. Chapters 3 to 10 also include Exercises so you can test your knowledge. If you think a chapter is too easy for you, skip directly to the Summary and Exercises to ensure you have not missed something. If a chapter is too hard, read it carefully, making sure you did not miss anything from previous chapters then take all the chapter's exercises to test yourself.

At the end of each chapter, the Further Reading section lists the most foundational and influential resources. These resources are highly recommended to the interested reader who would like to deepen his or her knowledge in specific areas of WSNs. However, they are not essential for understanding the chapters themselves.

WHAT DO YOU NEED TO WORK WITH THIS BOOK?

Obviously, you need the book itself. The book does not assume that you have any specialized experience with wireless sensor networks, but it does assume that you have some programming experience and basic hardware knowledge. Furthermore, it is highly advised to buy some sensor hardware, as discussed below. Apart from the hardware, you do not need any further financial investments as all recommended software tools are either open source or free for use.

Experience or knowledge in computer networking (e.g., TCP/IP, ISO model) is useful, but not necessary. Experience with wireless networks and their challenges are even more useful, but also not necessary. This book will teach you everything you need to know about wireless communications for sensor networks. At the same time, even if you are an expert of computer or wireless communications, you need to be aware of the fact that sensor networking is quite different than these.

PROGRAMMING PREREQUISITES

To benefit from this book, you need to be able to program in ANSI C (not C++). If you cannot do so, but you can program in another language, such as C++, Java, Python, or Perl, you will not find it difficult to learn ANSI C. If you have never programmed before, you should invest more time in learning C first and then return to this book.

There are many ways to learn or refresh your knowledge of C. There are online tutorials, books, and many mailing lists where you can find customized help. The best books to learn C are *Systems Programming with UNIX and C*, by Adam Hoover and *The C Programming Language*, by Brian Kernighan and Dennis Ritchie. Another good option to learn C is to take an online course.

The most important concepts you need from C are pointers and static memory management. This might sound like a step back into the middle ages if you are used to modern concepts such as garbage collectors and dynamic memory management. However, sensor nodes are too memory restricted to provide these functions so you need to allocate the memory often statically and to manage it manually. For this, you need to understand pointers. To find out whether you have sufficient C knowledge, please take the C quiz on the book's website.

SOFTWARE TOOLS AND THE CONTIKI OPERATING SYSTEM

All of the examples in this book and on the corresponding website are written in the Contiki operating system for wireless sensor networks. Contiki is open source and free for use both for non-commercial and commercial solutions. It is well documented and has an extensive community supporting it. For all of these reasons, it is ideal to learn wireless sensor networks, but also offers the possibility to directly use the developed solutions in any environment.

This book's website also provides tutorials to start working with Contiki.

SENSOR NODE HARDWARE

I strongly advise you to buy some sensor node hardware, at least two or three sensor nodes. Buying a single sensor node is not an option; it is like buying a single walkie-talkie. With two, you can let them talk to each other. With three or more, you can even build some interesting applications. If possible, try to get five nodes.

The Contiki website maintains an overview of supported hardware at their web-page: http://contiki-os.org/hardware.html.

Whether a specific platform is supported or not depends on the micro-controller and the radio used (see Chapter 2). A good option is the Z1 platform from Zolertia: http://zolertia.io/z1.

Z1 is popular in academia and the industry, and is fully supported by Contiki and its simulator Cooja. You can check this book's website to see whether this recommendation has changed, which other sensor nodes are supported, and where you can buy them.

Of course, sometimes it is not possible to buy sensor nodes. In this case, you have several options and this book will still be quite useful to you.

- Borrow from the local university. Almost every university in the world, which has an electrical engineering or computer science department, will also have a research group working in wireless sensor networks. You can typically find it in the computer networking or pervasive computing research areas. Thus you can ask the researchers whether you can borrow them for some time.
- Shared testbeds. The favorite testing tool of all WSN researchers is the testbed. A testbed is nothing more than sensor nodes, usually installed in a university building with cables connecting them to a central server and providing them with power. Shared testbeds also provide a web interface to program individual or all sensor nodes and to download experimental data later. An example of such a testbed is INDRYIA in Singapore.[1] If you really cannot find sensor nodes to work with, then a testbed is an option for some more advanced exercises to experience the hands-on feeling and properties of the real-world environment. However, a testbed remains a virtual environment, where you cannot see your

---

[1] indriya.comp.nus.edu.sg/

application running in real time. Furthermore, it is not trivial to prepare such experiments.
- Use a simulator. Contiki has its own simulator called Cooja. While this is a possibility, I do not recommend it because it will not offer you the experience and satisfaction of having something real in your hands. It is a little bit like learning a new language and being forbidden to speak it. However, Cooja makes it possible to use exactly the same programming code as for Contiki itself and is a good companion while debugging and experimenting.

HOW TO USE THIS BOOK: SUMMARY

You need programming experience in *ANSI C*, especially in concepts of pointers and static memory allocation. Before you start reading this book, you should complete the online quiz and consult the online references.

In terms of *software tools*, you need the Contiki operating system and its tools. You are also urged to look into the tutorials and installation guides, which are available online at this book's website.

In terms of *hardware*, you need at least two or three sensor nodes, although five is best. I recommend Z1 from Zolertia but others are listed on the book's website.

All necessary tools, tutorials and examples from this book along with updated information on supported hardware platforms are available on the book's official website:

  comnets.uni-bremen.de/wsn-book/

1

WHAT ARE WIRELESS SENSOR NETWORKS?

This chapter introduces wireless sensor networks, what are they as well as what types and applications exist. If you have previously worked with wireless sensor networks and know about their possible application areas, you may want to skip this chapter.

1.1 WIRELESS SENSOR NETWORKS

Wireless sensor network (WSN) is a collective term to specify a rather independent set of tiny computers with the main target of sensing some physical property of their environment such as vibration, humidity, or temperature. They consist of a few to thousands of *sensor nodes*, often also referred to as nodes or sensors, which **sensor nodes** are connected to each other via wireless communications. Typically, there is also at least one special node, called the *sink* or the *base station*, which connects the sen- **sink** sor network to the outside world. Figure 1.1 provides a general example of a sensor **base station** network.

There are several assumptions or general properties of WSNs, which make them different from other types of wireless networks.

The resources of individual sensor nodes are highly limited. In order to cover large areas for monitoring, the individual sensor nodes need to be cheap. In order to be cheap, their components need to be cheap. Thus, the absolute minimum is installed and used on sensor nodes so their hardware resembles more of a PC from

Introduction To Wireless Sensor Networks, First Edition. Anna Förster.
© 2016 The Institute of Electrical and Electronics Engineers, Inc. Published 2016 by John Wiley & Sons, Inc.

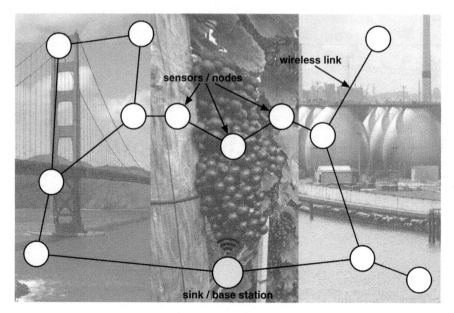

FIGURE 1.1   A typical sensor network with several sensor nodes and one base station. The sensors are connected to each other via wireless links, whereas the base station is typically more powerful and connected to the outside world. The application areas and environments are endless!

the 1980s than a modern device. All the properties and limitations of sensor networks come from this minimal hardware design. For example, this is the reason why not each of the sensor nodes can be equipped with a GPS receiver and a GPRS antenna for communication, but instead only one node can usually afford it (the sink/base station).

The wireless links are spontaneous and not planned. Different from other wireless networks, such as Wi-Fi hotspots, WSNs are not carefully planned to perfectly communicate and enable specific service quality levels. Instead, the assumption is that each of them tries to detect its brothers and sisters, and to exchange some minimally required data with them. Thus, WSNs are deployed (installed) quickly and without much knowledge of the environment. Existing experience with real WSNs and some theoretical foundations help installing more robust and self-sustainable networks than simply spreading them around the environment. However, the original dream of throwing sensor nodes out of an airplane to monitor thousands of square kilometers remains a dream.

The sensor network senses some phenomenon and needs to transfer the data to an external user. There is always something to sense out there: humidity, temperature, vibration, acceleration, sun radiation, rain, chemical substances, and many others. The main target of a sensor network is to sense some phenomenon and to transfer the gathered information to the interested user, typically an application residing somewhere outside the monitored area. The limited resources on the sensor nodes do not allow them to process the information extensively locally.

The main functionalities of a sensor node are *sensing* and *sending*.

1.2 SAMPLE APPLICATIONS AROUND THE WORLD

Vineyard monitoring is one of the most classical examples of sensor network monitoring. The goal is to reduce water irrigation and to predict or discover vine sicknesses as soon as possible. This not only minimizes costs of growing the vines through less water usage, but also enables organic growing with low usage of pesticides. Sensors used include air temperature, air humidity, solar radiation, air pressure, soil moisture, leaf moisture, ultraviolet radiation, pluviometer (rain sensor), and anemometer (wind sensor). The sensors are typically spread over a large area of the vineyard and deliver their information to an external database, in which the information is processed by special environmental models. The results are shown to the scientist or to the vineyard farmer and can be automatically connected to the irrigation system. Figure 1.2 shows

FIGURE 1.2   A sensor node from SmartVineyard Solutions installed in an organic vineyard in Slovakia. Reproduced with permission from SmartVineyard Solutions, Hungary.

a typical vine sensor node installed in a vineyard in Slovakia from SmartVineyard Solutions,[1] a Hungarian spin-off company.

A similar sensor network scenario is used for many other agricultural applications, often called precision agriculture. Examples are potato monitoring in Egypt [1], crop monitoring in Malawi [2], or a solution for vegetable monitoring on an organic farm in South Spain [3]. All of these systems have one problem in common: the foliage which develops over time. When the systems are first installed, the fields are almost empty or plants are small. However, as the crops grow, their foliage starts interfering with the system's work, particularly with its communications and sensors. Another common problem is that by harvesting time, the sensor nodes are well covered and hidden in the crops so their recovery is challenging. If unrecovered, they will most likely be damaged by the harvesting machines.

Bridge monitoring is a similar application in which the structural integrity of a bridge is monitored. Again, the space is limited, even if communication quality is better because of the outdoor, free-space environment. However, accessibility remains extremely limited.

There are two famous examples of bridges being monitored by sensor networks. The first example is tragic. On August 1, 2007, a bridge spanning the Mississippi river in Minneapolis collapsed suddenly under the weight of the rush hour traffic, killing 13 people and injuring another 145. The bridge was rebuilt shortly thereafter, this time equipped with hundreds of sensors to monitor its health and give early warnings.

The second example is more positive and presents the six-lane Charilaos Trikoupis bridge in Greece, which spans the Gulf of Corinth (Figure 1.3). It opened in 2003, with a monitoring system of more than 300 sensor nodes equipped with 3D accelerometers, tilt meters, tensiomag sensors, and many others. Shortly after opening, the sensor network signaled abnormal vibration of the construction's cables, which forced the engineers to install additional weights for stabilization. Since then the bridge has not had any further problems.

Fire detection is crucial to save life and prevent damages. Sensor networks can be efficiently employed to detect fires early and send an alarm with the fire's exact position to fire brigades. This idea has been used worldwide in countries such as Spain, Greece, Australia, and Turkey. The main challenge with this application is the sensor node hardware itself and its resistance to high temperatures. It is quite inefficient if the sensor network designed to detect fires fails on a hot day or at the first sparks of a fire. For these reasons, robustness of the hardware and smart, over-provisioned network design are essential. Furthermore, typically large remote areas have to be covered, which can pose problems for installation, maintenance, and communication.

Tunnel monitoring is another way in which sensor networks are used. Road tunnels are dangerous all over the world. If something happens, such as an accident or a fire, it is essential to know how many people are still inside and exactly where they are located. Even in normal situations, tunnels are a danger by themselves, as the light

---

[1]http://smartvineyard.com

FIGURE 1.3   The Charilaos Trikoupis bridge in Greece, with a sensor network installed with over 300 sensor nodes. In its first days, it signaled abnormal vibrations, which could be quickly fixed. Source: Guillaume Piolle, Flickr.

outside often blinds the driver when exiting the tunnel. The TRITon project[2] in Trento, Italy is an example of how to resolve these problems with sensor networks. Light sensors are used to adaptively regulate the light inside the tunnel to avoid both energy wasting and driver blinding. Video cameras are used to automatically detect cars and track them until leaving the tunnel for rescue operations. One of the main challenges of tunnel monitoring is the shape of the tunnel because wireless signals scatter significantly in such narrow environments and communications become obstructed and unreliable. Additionally, there is not much space to install many nodes and to make the network more robust. Working in tunnels is also difficult because the tunnel needs to be closed and guarded during work.

Animal monitoring is one of the oldest applications for sensor networks. These are not typical sensor networks because the main phenomena they sense are the movement of the animals. Thus, they function more as tracking applications, implemented with typical sensor nodes. Figure 1.4 shows how an Australian farm uses network sensors for cattle farming [4]. Instead of installing large area electric or other types of real fences, the cattle are equipped with a collar that has a positioning system with speakers and an electro-shocker. A web interface allows for drawing fences on a

---

[2]http://triton.disi.unitn.it

FIGURE 1.4   Virtual fences sensor network for cattle farms in Australia. Source: CSIRO, Australia.

virtual map, identifying where the cattle should or should not be. If a cow tries to cross the boundary to a locked area, it first receives an acoustic signal. If that does not work and the cow stays in the prohibited area, a light electro shock is given (similar to the ones given by electric fences). The cows quickly remember that the acoustic signal is followed by the electric one and learn to react quickly and leave the locked area.

Additionally, the movement of the cattle can be used to evaluate their fitness. Other sensors can even monitor their temperature which allows for further health monitoring. This is especially important for large area cattle farms such as those found in Australia or South America.

Animal recognition is yet another area of animal monitoring. One of the first applications of wireless sensor networks was developed in Australia to search for the cane toad, a special toad species. The history of the cane toad is worth knowing. The cane toad is native to South and Central America but was introduced in Australia in 1935 to fight grey backed beetles, a species threatening the cane fields. However, this introduction was not very successful since grey beetles live on the tops of the cane but the cane toad is not a very good climber. Thus, the cane toad ignored the beetles and instead began eating other native species such as lizards. The cane toad population grew to such an extent that local rangers purposefully began killing them, yet they hid well.

In this case, the uniqueness of the sensor network is its sensor. It literally listens to the forest's sounds to recognize the specific quaking of a cane toad. Once recognized, an event is generated and sent along with the cane toad's coordinates to local rangers who then can physically find it. This was a significant increase in efficiency because previously the rangers did the work of the sensor network and went to listen for the cane toad's quaking.

Technically, the sensor network is not trivial to implement. It requires on-board audio processing to identify the cane toad quaking, which also requires sufficient on-board resources. The full system was described several years after the deployment by Hu et al. [5].

There are many different applications of wireless sensor networks around the world making it impossible to cover all their application areas here in this book. The interested reader should simply search the Internet for further applications and examples of wireless sensor networks as most of them are well described.

1.3 TYPES OF WIRELESS SENSOR NETWORKS

Recently, various new definitions and buzzwords have emerged in the area of sensor networks. This section explores them in more detail.

Wireless sensor network is still one of the main terms for describing a network of cooperating tiny computers that are able to sense some phenomenon in the environment and send their results to a central entity such as a database or a server. The term is often used for installations, which work in the traditional way of sense and send. However, it can be seen also in broader contexts, depending on who is using the term and where.

Cyber-physical system (CPS) is a newer term for a wireless sensor network. It attempts to better describe what you can actually do with these networks and their main properties when being integrated into a physical environment. Different from other computers and devices, which are environment agnostic, cyber-physical systems are part of the environment and application restricted. Another important property is the fact that they also can affect the environment via so-called *actuators* such as in automatic irrigation pumps, light switches, alarms, and humidity or temperature regulators. Interestingly, the term wireless is no longer used in cyber-physical systems. The reason being, many sensor networks are now wired, since power supply in buildings is easier to maintain through wired connections than through wireless ones. Smart homes are a typical example of cyber-physical systems (Figure 1.5) as are smart grids and smart cities.

actuators

Body sensor networks refer to a specific type of network, designed to be carried on the body (mostly human). Applications include health monitoring, weight

SMART HOME

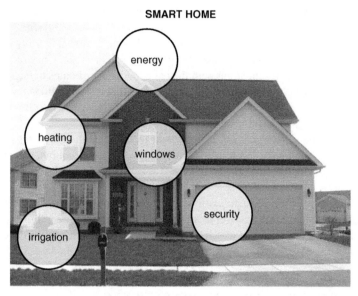

FIGURE 1.5   A typical example of a cyber-physical system is the smart home application in which various sensors and actuators are combined for the convenience of the users.

management, sports logging, and many others. There are some peculiar examples, such as smart shoes or smart T-shirts, which are able to sense your activity or heart rate. Most of the sensor nodes are tiny, sometimes even implantable. The current trend is clearly moving towards one integrated device that can sense all functions, instead of several different sensors on various parts of the body. The whole sensor network often degrades into one or very few devices, such that the term network becomes meaningless. Figure 1.6 presents an example of body sensing.

Participatory sensing, collaborative sensing, or crowdsourcing refers to a new and fast developing type of sensing in which sensors are essentially humans with their smartphones. For example, people can track their biking paths and then evaluate them in terms of safety, noise, or road quality. All this data are gathered on a central point and processed into a single biking quality map of a city, which can be distributed to any interested user [6]. The real power of these applications is that no additional hardware is needed, only a rather simple user-oriented application for smart phones. At the same time, this is also the largest challenge in these applications—motivating people to join the sensing process and to deliver high quality data. Privacy is a big issue with these applications because it requires people to share their location data.

The Internet of Things (IoT) is often mistaken to be a sensor network. However, IoT's main concept is that all things, such as a washing machine or radio, are

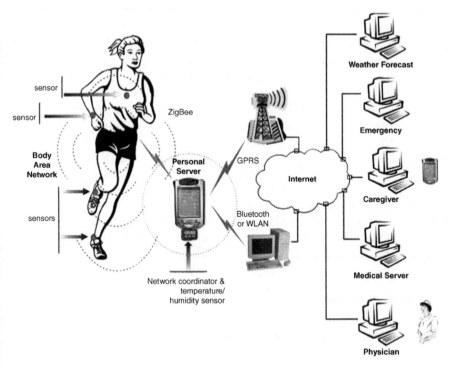

FIGURE 1.6   A body sensor network consists of relatively few sensors on or in the body of a human. Source: Wikipedia.

connected to the Internet. The Internet connection has significant advantages when implementing sensor networks and can thus be seen as an enabling technology. However, the target can also be very different, e.g., when you can read your emails from the microwave or from your car. The term Internet is also a hint that usually these networks are IP enabled and thus use a well-defined communication stack. This can be seen as an advantage (no need to reimplement) or as a disadvantage (high energy use, little flexibility).

Figures 1.5 and 1.6 present some of the preceding network types, a cyber-physical system for a smart home application and a body area network for health monitoring. The common property of all these applications and types is that the system is highly integrated with its respective environment and cannot be easily transferred or generalized.

Regardless of the exact term used—wireless sensor networks, sensor networks, body networks, or cyber-physical systems—the most important topics remain *sensing, communication, energy management*, and *installation*.

SUMMARY

Wireless sensor networks are a modern technology covering the general application area of monitoring. They typically consist of many (few to thousands) devices to enable large-scale monitoring. The main requirements of WSNs are to be *small-size* and *low-cost*. Thus, instead of implementing complex applications in the network itself, the sensor network performs two main operations: *sensing* and *communicating* the sensed data to each other and to a central server.

Applications of wireless sensor networks span environmental and animal monitoring, factory and industrial monitoring, agricultural monitoring and automation, health monitoring and many other areas. One of the main characteristics of wireless sensor networks is that they are tightly *coupled with their application*. A system developed for agricultural monitoring in one place can hardly be used in another place and is almost impossible to use in animal monitoring.

FURTHER READING

All of the readings presented in this chapter refer to detailed descriptions of a particular sensor network application.

[1] Sherine M. Abd El-kader and Basma M. Mohammad El-Basioni. Precision Farming Solution in Egypt Using the Wireless Sensor Network Technology. *Egyptian Informatics Journal*, **14**(3):221–233, 2013. ISSN 1110-8665.

[2] M. Mafuta, M. Zennaro, A Bagula, G. Ault, H. Gombachika, and T. Chadza. Successful deployment of a Wireless Sensor Network for precision agriculture in Malawi. In *IEEE 3rd International Conference on Networked Embedded Systems for Every Application (NESEA)*, pages 1–7, Dec 2012.

[3] J. A. Lopez Riquelmea, F. Sotoa, J. Suardiaza, P. Sancheza, A. Iborraa, and J. A. Verab. Wireless Sensor Networks for precision horticulture in Southern Spain. *Computers and Electronics in Agriculture*, **68**(1), 2009.

[4] Swain. D., G. Bishop-Hurley, and J. Griffiths. Automatic cattle control systems -grazing without boundaries. *Farming Ahead*, June 2009.

[5] W. Hu, N. Bulusu, C.T. Chou, S. Jha, A. Taylor, and V.N. Tran. Design and Evaluation of a Hybrid Sensor Network for Cane Toad Monitoring. *ACM Transactions on Sensor Networks*, **5**(1), 2009.

[6] S. Verstockt, V. Slavkovikj, P. De Potter, and R. Van de Walle. Collaborative bike sensing for automatic geographic enrichment: Geoannotation of road/terrain type by multimodal bike sensing. *Signal Processing Magazine, IEEE*, **31**(5):101–111, Sept 2014. ISSN 1053-5888.

2

ANATOMY OF A SENSOR NODE

This chapter explores the general concept of sensor nodes and some examples. There are a great variety of sensor nodes ranging from tiny implantable body sensor nodes to smartphones and tablets. If you are a computer scientist or experienced in operating system concepts and energy management of hardware, you can skip directly to the summary of this chapter. The summary will provide you with the main differences between sensor nodes and more traditional IT systems.

2.1 HARDWARE COMPONENTS

The typical components of a sensor node include the following:

- **Microcontroller.** This is a small computer, which already includes a processor, some memory, and general purpose input/output ports, which can be configured to connect to various external devices (radio, sensors, etc.). It does all of the computation on board the sensor nodes and coordinates the work of all external components.
- **Radio transceiver.** A component for wireless radio communication. This typically works in both directions (receiving and sending) and is one of the most energy hungry components on board. Very often the radio transceiver also has an internal microcontroller, which allows it to buffer packets, validate them, and implement some low-level protocols, such as IEEE 802.15.4 (part of the Zigbee standard) or Bluetooth.

Introduction To Wireless Sensor Networks, First Edition. Anna Förster.
© 2016 The Institute of Electrical and Electronics Engineers, Inc. Published 2016 by John Wiley & Sons, Inc.

- **Sensor.** Some node platforms do not come with sensors on board because they are targeted to communication development only. However, most of the existing platforms come at least with some sensors, such as temperature, humidity, light, etc. Other, more expensive and application specific sensors, such as wind or rain sensors, can be found on industrial-oriented platforms.

- **External memory.** Very often you can also find some external memory either on board (like on the Z1 platform, which has 16 MB external non-volatile memory) or an optional slot such as a MicroSD. This allows the possibility to permanently store some larger amounts of data instead of always streaming it to a sink/base station.

- **Battery.** Every sensor node needs energy to run. Sometimes it is one or two standard battery slots such as AAA or AA. Other times it is a rechargeable battery; sometimes even a solar rechargeable battery or an energy harvester.

- **Serial adapter.** Almost every sensor node comes with some sort of a serial communication, usually a cable which can connect the sensor node to some other device (most often to your PC). The serial adapter is used for programming and debugging the node and sometimes for charging it. Z1 has a microUSB connection for power supply and debugging.

Figure 2.1 shows all the hardware components of the Z1 platform used in this book. You can see the microUSB connector, temperature sensors, accelerometer, connectors for additional components (called Phigdets for Zolertia products), and the embedded antenna of the radio.

embedded antenna   An *embedded antenna* is useful because it renders the node smaller but still enables rather good communication. However, an embedded antenna is never as powerful and homogeneous as a normal antenna. Thus, the node provides a connector for an external antenna.

oscillator   The *oscillator* enables the sensor node to track time (not visible in Figure 2.1). However, all it can do is count the ticks after the node's last reboot. There is no such

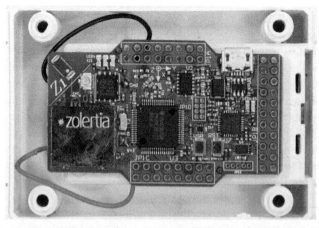

FIGURE 2.1   Z1 sensor node platform. Source: Special thanks to Jens Dede, University of Bremen.

thing as real global time on sensor nodes. Moreover, the used oscillators are cheap and not very reliable because they tend to drift away from real time. Chapter 7 explores this issue in greater detail.

2.2 POWER CONSUMPTION

One of the most important properties to understand about sensor node hardware is its power consumption. Each of the components of a sensor node requires energy to operate. This energy is highly restricted and needs to be provided by on-board batteries. Thus, it is crucial to understand which components are most energy hungry and to use them only when absolutely needed.

Table 2.1 gives an overview of the power consumption of the most important components of the sensor node platform Z1. The following discussion focuses on the targeted platform used in this book, but also covers most of the currently available hardware.

Let us first understand how to read this table. All of the power consumption is given in Ampere (A). For example, when the radio is sleeping and off (it cannot send or receive anything), it still consumes 0.01 mA or 0.00001 A. The batteries attached to the sensor node are expected to provide a constant voltage of 3 V (if there are two batteries, each supplying 1.5 V). Each battery has some initial (*nominal*) capacity, measured typically in Ampere-hour (Ah). For example, a typical AA battery has a *nominal capacity* of 2800 mAh. Thus, in theory, if you draw 0.01 mA from this battery continuously, you should be able to survive for 2800 mAh/0.01 mA = 280,000 h, which is approximately 32 years. However, that is only considering the radio's sleep mode and a constantly sleeping radio is not very useful. If you assume that both the radio and the microcontroller are on and working, they have a total of 1.8 mA + 19.7 mA = 21.5 mA and the battery will survive for 2800 mAh/21.5 mA = 130.23 h, which is approximately five and a half days.

nominal battery capacity

TABLE 2.1   Nominal power consumption of Z1 node components. Data from www.zolertia.io, www.ti.com, and www.micron.com.

Component	Mode	Current Draw
Microcontroller (TI MSP430)	Active	1.8 mA
	Sleep	5.1 μA
RF Transceiver (CC2420)	Receive	19.7 mA
	Transmit (at 0 dBm)	17.4 mA
	Sleep	0.01 mA
Accelerometer (ADXL345)	Standby	0.0001 mA
	Active	0.04 – 0.145 mA
External flash (Micron M25P16)	Write	15 mA
	Read	4 mA
	Sleep	0.001 mA
Temperature sensor (TMP102)	Sense	0.015 mA
	Sleep	0.001 mA

The preceding calculations are strictly theoretical and real batteries or hardware may behave differently. However, this also illustrates how important it is to minimize the usage of individual components. Even if batteries do not behave like perfect containers of energy and, for example, degrade with time, there is still a huge difference whether you draw 0.01 mA from them or 21.5 mA.

In fact, the preceding calculations are not that far from reality, when all of the components are always on. In experiments, the lifetime of sensor nodes with batteries were measured with the microcontroller and the radio on the whole time [3]. The measurements were taken on a TelosB (www.memsic.com) node, which is very similar to the Z1. By taking into account only these values, a theoretical node lifetime of approximately 140 hours was calculated. In experiments, it was measured to be only approximately 80 hours. The difference is substantial, but not overwhelming and could be explained by the used batteries' properties, calculations taking into account only the main components, etc. Figure 2.2 compares the experimentally obtained values with theoretically calculated ones.

When introducing sleeping times for the radio (the so-called duty cycle), the picture changes dramatically. In this scenario, the radio is switched on and off periodically to enable power saving. The difference between calculated and experimental values grows exponentially (Figure 2.2). For example, with a duty cycle of 25%, which means that the sensor remains in sleep mode 75% of its lifetime and in an active mode for the remaining part, the difference between reality and theory is 310 hours or 260%.

Let us now turn back to Table 2.1. The microcontroller seems to be drawing only 1.8 mA at most, so further discussion here is not needed. However, it should be obvious that to avoid any unnecessary work we need to put the microcontroller to sleep

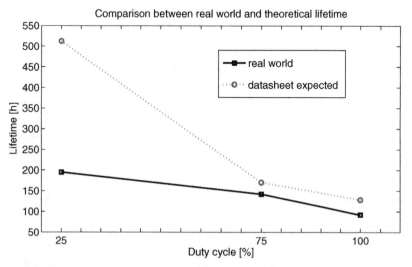

FIGURE 2.2   A comparison between the data sheet calculated sensor node lifetimes and experimentally confirmed ones.

as often as possible. The next component is the radio. It is interesting to note that not only is it power hungry, but that receiving packets need more power than sending packets. This might not seem logical at first glance. However, it is "easier" to send out packets than to receive some, because receiving needs to filter out noise and perform other functions that are not necessary for sending. This is similar to talking and listening in a crowded, noisy room. Talking is relatively easy, as you just talk normally. When trying to listen, you need to concentrate on only the person you want to listen to and filter out everything else.

Let us now look in more detail at what these new insights mean for your practical work. As mentioned, a battery is supposed to provide a constant nominal voltage of 1.5 V. However, this works only in theory. In practice, the voltage starts dropping at some point. How the discharge drops is called the *discharge curve*. The problem **discharge curve** is that there is not a negligible time when the batteries still provide enough voltage for the components to do something, but not enough to do it properly. For example, the radio is still able to send something out, but not to receive anything [3]. Some components stop functioning while others continue. For example, the flash memory requires quite a high voltage to do something, whereas the microcontroller survives at very low voltages. You can find many more details on batteries on the website of Battery University.[1]

Another peculiarity are the sensors themselves. They actually depend on some reference voltage to correctly sense their phenomenon (e.g., temperature). If this reference voltage cannot be achieved because the batteries cannot supply it, the sensor continues delivering values, but they are wrong. This is experimentally shown in Figure 2.3, where the input voltage to the sensor node has been gradually decremented, while keeping the temperature of the environment constant. You can clearly see that at approximately 2.05 to 2.04 V, the temperature sensor starts delivering faulty results, which change exponentially fast.

These last "breaths" of the sensor node are very tricky to identify and to manage. The simplest way to do it is to sense the supply voltage (an internal sensor in almost all microcontrollers) and once below a certain minimum threshold to shut down the node.

2.3 OPERATING SYSTEMS AND CONCEPTS

This book uses the Contiki operating system, which is one of the most-used sensor network operating systems for education and research. It is a well supported and quickly developing project that regularly issues new versions. This chapter focuses on some long-lasting and general operating system concepts instead of going into implementation or usage details. Those are detailed and regularly maintained on this book's web site. There are also other operating systems, both open source and proprietary. The interested reader can turn to the sensor networks programming survey of Mottola *et al.* [2].

---

[1]batteryuniversity.com

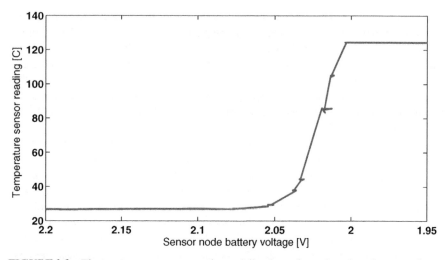

FIGURE 2.3   The temperature sensor continues delivering values when the reference voltage drops, but they are faulty. Source: Special thanks to Thomas Menzel for these measurements.

What is an operating system? An operating system is software, which manages software and hardware resources on a computer system and provides services to application software. The interaction of all components is shown in Figure 2.4. The main resources of the system are typically the CPU, the memory and input-output devices. When more than one application runs on a system, it is necessary to somehow split the usage of these resources among the applications. For example, it is clear that two applications cannot use the CPU exactly at the same time—they need to take turns. **CPU scheduling** This is called *scheduling* and it ensures that lengthy processes or applications do not overtake all of the computation resources. In the case of sharing memory, the operating system needs to make sure that one application does not change the memory of

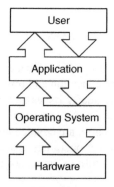

FIGURE 2.4   A general representation of interactions between the platform hardware, operating system, applications, and the user.

another one. The concept of reserving memory for each individual application (often referred to as process) is part of the so-called *multitasking*. Another very important aspect of operating systems is how they manage external devices, such as keyboards, printers or in this book's case, sensors, buttons, or the radio. The following section discusses these concepts.

multitasking

2.3.1 Memory Management

Memory management refers to how available memory is managed on a system. Imagine you have a total of 1 KB = 1,024 bytes of memory on a sensor node (in the microcontroller). This is all the memory the software can use to store variables, buffer and process data, etc.

This type of memory is typically also called *volatile memory*, because it is erased when the node is rebooted. It is also *RAM* type memory (random access memory) because it can be accessed at any point.

volatile memory

RAM

In order to use this memory, you need to allocate parts of it to variables. Let us say, in your 1 KB memory, you would like to allocate some memory to a variable called COUNTER. You can do this either *statically*, before the application even starts, or *dynamically*, when the application has already started. What is the difference? The greatest advantage of dynamic memory allocation is that you do not have to decide at coding time how much memory you will need for what. For example, imagine you are storing in COUNTER the incoming packets. How many could arrive together? If you allocate the memory statically, you need to decide at coding time what the maximum number of incoming packets is that you need to store. If more come, you would need to delete those extra packets. If less come, some of the memory will remain empty and unused. If you allocate the memory dynamically, i.e., during application run, you could start with COUNTER occupying zero bytes and then extend it every time a packet is received.

static memory allocation

dynamic memory allocation

However, static memory also has advantages. When the total memory is severely restricted (like in almost all sensor node platforms), it is risky to use dynamic memory because you might easily exceed the memory you actually have. This results in a memory allocation error, which either crashes the application or at least blocks its normal work. With statically allocated memory, you can calculate at coding time how much you can use for what.

Memory Management Implementation.   Yet another problem of dynamic memory allocation is its actual implementation. Allocating static variables in the memory is easy, as the compiler just orders all of them one after another in the available memory. However, dynamic memory management is tricky. You are allowed to allocate and again erase variables with various sizes. Let us assume a very limited (and not very realistic) memory of 20 bytes total, as shown in the following example:

start - all empty

Imagine you have first allocated a COUNTER variable of 2 bytes. The easiest way is to allocate it in the very beginning of the memory (imagine simply the memory as a chain of bytes):

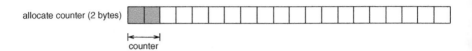

Next, allocate a new variable of 8 bytes called DATA:

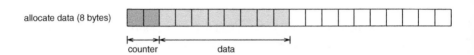

Now, you have 10 bytes out of 20 allocated. Perhaps you decide that you do not need the COUNTER variable anymore and delete it:

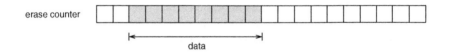

memory fragmentation

At this point, you have the first 2 bytes empty, 8 bytes allocated, then 10 bytes empty again. When this occurs, the memory is defined as *fragmented*. This is a tough problem because you cannot easily accommodate large portions of memory in fragmented memory. Now, the dynamic memory management has two options: either reallocate (shuffle) the allocated variables or wait until other memory allocations make this necessary. For example, next you might decide to allocate a variable PACKET with 12 bytes. If the memory remains unchanged after erasing COUNTER, you do not have a chunk of 12 bytes anymore. The dynamic memory management has to now split the new variable into two chunks or reshuffle everything:

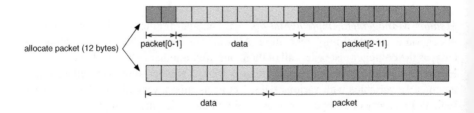

memory defragmentation

The second option is called *memory defragmentation*. Both options require lengthy processing and are memory-intensive.

Thus, we can conclude that:

> **Static memory management** is simple and does not need any processing or memory by itself. However, it is also inflexible as it cannot be changed once the application is loaded on the sensor node. **Dynamic memory management** offers much more flexibility, but can easily overflow in memory-restricted devices and requires memory and processing overhead by itself.

This is the reason why many sensor node operating systems do not offer dynamic memory management at all or only with limited capabilities. Instead, it is assumed that the programmer is taking care of his or her variables and reuses them as often as possible instead of deleting them.

Contiki supports both options, but dynamic memory should be used with caution. Please refer to this book's website for more information about currently supported memory management and its usage. As mentioned in the introduction of this book, such details change too often and too dramatically to be reflected well in a long-lasting medium like a book hence we maintain them only on the web site.

Example of Memory Management.   This example explores in more depth how to manage available memory on your own. It will teach you two things: how to implement efficient algorithms for resource-restricted devices such as sensor nodes and how to manage memory wisely.

Let us assume you need to store packets in the system. However, there are different types of packets such as incoming packets and outgoing packets. Sometimes an incoming packet might become outgoing, if you need, for example, to resend it. What you need is a way to manage these packets using the least amount of processing possible and to efficiently use the available memory.

Whatever implementation you decide on, you need to explore it in terms of performance when conducting the following operations:

- Moving packets from one type to another.
- Finding a currently free packet to use.
- Deleting a packet (contents).
- Finding the next packet to send out or to process.
- Resolving how many packets from each type you can use.

How can you evaluate the performance of such operations? The main question is how much time or how many simple operations (e.g., add or copy a byte) do they take to finish? For example, if you want to copy N bytes, you would need N copy operations. If N starts growing, the number of copy operations will also grow linearly with N. Other operations, such as matrix multiplication, take a non-linear number of operations to complete. What is important to you is the approximated time for completion, not the exact number of operations. This is because when the input grows quite large, it does not matter whether you need N, 2N, or 3N + 2000 operations. However, it does matter whether you need N, N^2, or even 2^N operations.

To express this dependency, we use the so called *big-O notation*. We will not explain the mathematical background of this notation here but the interested reader can turn to the book of Cormen *et al.* [1]. However, its meaning is important: when a procedure or an algorithm takes O(N) steps, this means that the real number of steps needed to complete this procedure is linearly proportional to N, where N is the size of the input to the operation. Some operations have O(1) complexity, because they always need a constant number of operations to complete, independently of how large the input is. Other operations, such as the traveling salesman problem, take $O(c^N)$ steps or even O(N!) steps to complete. Coming up next, we will be using the big-O notation to evaluate the performance of our memory management algorithms.

Trivial implementation with static arrays. A first and trivial implementation could allocate a static number of packets to each type. For example, you could allocate 5 packets to each type, which is depicted in Figure 2.5.

First, if you receive more packets at once than you can store in the INCOMING array, you need to drop any additional ones. The same is true if you need to send out more than 5 packets. Furthermore, if you need to move one packet from one array to the other (because you need to send out a packet again, which just came in, for example), you need to copy its contents into a packet from the other array. This is time and processing consuming. To be exact, it takes exactly L copy commands, if L is the number of bytes in a packet. This number might be slightly different, depending on the real implementation of the copy command.

Thus, your copy operation within your trivial memory management implementation takes O(L) steps, where L is the length of a single packet.

Let us now explore how fast you can find an empty packet in any of the two arrays. You need to go through the array and see whether the packets are empty or not. In the worst case, this will take exactly N steps, where N is the size of the array. This happens when no packets are empty and you arrive at the end of the array without

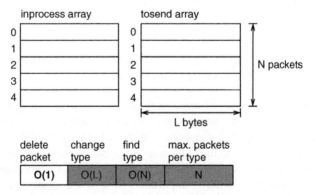

FIGURE 2.5 A trivial memory management implementation.

finding a free one. A search for an empty packet takes O(N) steps, where N is the size of the array.

How many steps does it take to find the next packet to process or to send out (a non-empty packet)? The answer is the same as for empty packets, at most N steps, or O(N).

How many steps does it take to delete the contents of a packet? In fact, it is not necessary to delete all of the contents, it is enough to mark one of the fields as illegal (e.g., −1 for the ID of a packet or something similar). Thus, this operation takes a constant number of steps, independent of the size of the packet or the size of the array. Thus, it takes O(1) steps.

The summary of the performance analysis is provided in Figure 2.5.

Shared memory organized by type. The biggest problem of the previous trivial implementation was the number of packets you can handle. It often happens that many packets arrive at the same time or that you need to create many packets at once. In this case, you have to allocate a lot of memory to not drop any packets but you will rarely use it.

Another possibility to implement memory management is to share the memory between both types of packets. Instead of allocating two arrays, you could allocate one large array and use it for both types of packets. Of course, you would need to differentiate between the types of the packets, for example by adding an additional field to each of them. It would be even more convenient to use this field not only to identify your two types of packets, but also the empty packets. An example of this implementation is provided in Figure 2.6 and the next discussion explores its performance.

Moving packets from one type to another is simple. You need only to change the type ID in the packet, thus you need O(1) steps. Deleting a packet is also very easy as you just change the type to empty, again, O(1).

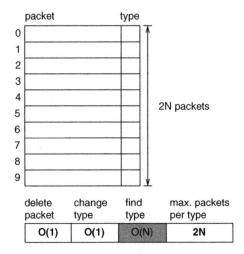

FIGURE 2.6   A shared memory management implementation.

Finding a packet of some specific type (whether empty or not) will take in the worst case as many steps as the whole array is large. Assume that you have taken an array exactly twice as large as the individual arrays from your trivial implementation. Thus, you use exactly the same amount of memory as in your trivial implementation and you can easily compare. In big-O notation, this is still O(N), as you ignore the constant of 2.

The last question is, how many packets from each type can you use? In this case, you can use as many packets as are free. In the best case, these are all 2N packets. This is twice as many as in your trivial implementation and it will make a big difference in a real application.

The summary of the performance analysis is given in Figure 2.6. The performance already looks better, as you can accommodate twice as many packets in the same amount of memory. However, you still need too much time to search for empty packets or next packets to process. To solve this issue, you will use something called **linked lists**   *linked lists*.

Linked list implementation. A linked list is a list of elements, in which each one shows to the next element. For this, each of the elements has an additional field, typically called NEXT, which is a pointer to the next element. In your case, you can use the type fields from your previous implementation for the NEXT fields. An example is provided in Figure 2.7.

How do you manage such memory? You would also need some initial pointers, which tell you where to start looking for a specific type of packets. For example, the initial pointer empty points to element number 2 in the array of packets. This means that you can find the first empty packet exactly there. The NEXT field of element 2 is supposed to point to the next empty packet, if any. In this example, it points to 5. Again, the NEXT field of element 5 should point to the next empty packet. In this case, it says −1. This means that there are no more empty packets and the linked list ends here. Thus, you have a chain of empty packets, which can be easily followed

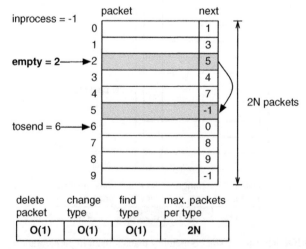

FIGURE 2.7   A linked list memory management implementation.

one-by-one until it finishes. For each type of packet you want to have, you would need exactly one initial pointer. What is the performance of such a linked list?

Finding the next empty or non-empty packet becomes trivial because the initial pointers always point exactly to those. Changing the type of a packet is more tricky. What you need to do is to change the chain of the next fields to point differently. This change is strictly localized to the packet, whose type you are going to change, its neighbors in its current list, and the initial pointer of the list where you want to put it. Thus, the number of steps you need are O(1), because they do not depend on the length of the packets or on the size of the shared array.

Deleting a packet does not exist anymore, as you need simply to move it to the empty list (as in type organized implementation).

How many packets can you use for each type? The answer is the same as with type organized implementation. Given that you still have a 2N sized shared array, you can use all those 2N packets for any type.

The linked list implementation is typically preferred, as it offers very good performance with very little memory.

> The way one solves a particular implementation problem has a large impact on the efficiency and thus on the energy expenditure of a system. Recall also the power consumption data of the individual hardware components earlier in this chapter in Table 2.1. As the user of a particular system, you cannot change much in terms of what the hardware requires in order to run. However, you do have great influence on how you use the hardware.

2.3.2   Interrupts

Interrupts are actually quite well defined by their own name—they interrupt the currently working program to do something else. For example, imagine you have an infrared sensor able to spot moving objects such as people passing. How do you check when somebody has passed? One possibility is to continuously check the sensor's input in an endless loop. However, that would block the processor with useless operations, rarely delivering a result. Another possibility is to implement an endless loop, which also checks all sensors one by one, does some blinking or processing, etc. However, this solution is also rather wasteful.

In result, interrupts were introduced. When an interrupt happens (an input bit to the microcontroller changes), the normal work is interrupted and a special handling function is called, which is implemented by the programmer. Once the function finishes, the microcontroller continues exactly where it was interrupted (Figure 2.8). The program is waiting for sensor data to arrive. The first possibility (left in the figure) is without interrupt handling, thus the program needs to check continuously and actively whether the data is there. In between, it can do something more useful, too, but it needs to check regularly. With interrupts, the program is free to do the useful work continuously. When the interrupt arrives, this work will be interrupted and later restored automatically.

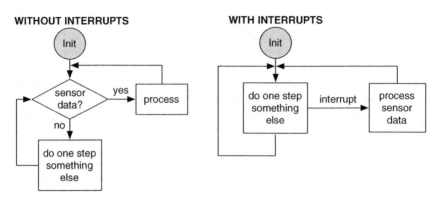

FIGURE 2.8   Waiting for sensor data to arrive without and with interrupts.

Interrupts are extremely useful for sensor networks, as you typically have a lot of sensors and/or actuators connected to the microcontroller. However, events happen rather rarely and the microcontroller can remain in power-saving mode until an interrupt arrives.

Events are often mistaken for interrupts. Their definition is actually quite similar in that an event is something that happens asynchronously (it cannot be predicted when it will happen). For example, an event is when the user pushes the button or when the sensor has a new value. However, the implementation of interrupts and events is different. Interrupts are implemented at the lowest level (the so-called instruction set level). Events are implemented as a waiting loop, exactly as was previously described. Events are typically used in higher application layers and many programming frameworks rely on software-defined events, e.g., the event based simulator OMNeT++ or the Contiki operating system.

2.3.3   Tasks, Threads, and Events

Continuing discussion on interrupts, one of the main problems in sensor network operating systems is how to organize access to resources and how to make those work efficiently. In order to better understand why this is a problem for sensor networks, let us first look at what the options are.

multitasking   Modern operating systems use *multitasking concepts*. A task is an independent process in the system, which has its own memory allocated and works completely independently from the other tasks. In this case, all tasks together are said to be working simultaneously. What happens in reality is that a hardware component, the so-called Memory Management Unit (MMU) makes sure that the memory is allocated and that different tasks cannot corrupt the memory of each other. Then a scheduler (software component of the operating system) is used to orchestrate the different tasks and give all of them some time to progress, before switching to the next one.

Since each of the tasks has its own memory, there is no problem to interrupt the task somewhere and continue again later. When accessing shared resources (like writing to the hard-disk or using the radio), the tasks are allowed to block the system for some time by defining *atomic operations*. These need to run until completion so that shared memory does not get corrupted. Atomic operations are also used for other situations such as when something could go wrong if they were to be interrupted.

atomic operations

Multitasking has one big advantage. It is very easy to define the individual tasks and to program them. However, they also have one big disadvantage. Most microcontrollers do not have a MMU because it is expensive and power hungry.

Thus, many sensor networks operating systems have turned to so-called *event based programming*. An event, as previously explained, is different from an interrupt as it typically comes from software and is not real time. For example, when you press the button of the mouse, you would like to generate an interrupt to see the result of the button press immediately. However, if your software is ready with some computation, it creates an event, which can be processed some time later by the waiting application. It has to happen soon, but not immediately. This is also the reason why interrupts are implemented at a low level whereas events are at a higher level.

event based programming

An event is a message that something has happened. Each event can be tied to some function, which processes the event. Such functions are called *event handlers*. Once an event process has started, it always runs to completion. For example, if your application signals that some sensor data is now available, the processing of this event (creating a packet, putting the sensor data inside, sending out the packet) will run to completion before the next event is processed. A scheduler again takes the role of ordering the events one after another. Interrupts have priority, but do not interrupt (or *preempt*) the current event processing. The trouble with event based programming is its high complexity. The programmer needs to first design a *finite state machine* and to follow this closely when implementing the application. Introducing one new event means to make sure that any sequence of events is still valid and correct. The following discusses an example of what a finite state machine is.

event handlers

preempting finite state machine

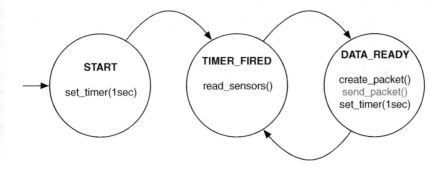

A finite state machine is a set of states in which the system can be currently. In this case, you have the states: START, TIMER_FIRED, and DATA_READY. The arrow going into the START event means that this is the first event after booting the system. The scheduler orders the events according to their arrival. Here you can assume that the timer is long enough to allow the radio to send out the packet before the next data is available. It is the programmer's responsibility to make sure that the sequence of events is correct. The following reflects how it looks when you want to make sure that the packet has been sent before you even start the timer.

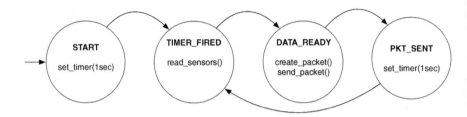

It becomes clear that even simple programs require a large number of events and it is hard to keep track of them and make sure all is correct and safe. Event based programming is more memory efficient because the memory is shared. However, it is quite complex for programmers. Additionally, think what will happen if one of the events requires a lengthy computation. It will completely block the system, since its event handler needs to run to completion. In the meantime, many different packets may arrive and their events will need to wait. The packet buffer may even overflow and some of them could get lost forever. This is a difficult problem for event-based operating systems.

blocking   Another important concept is *blocking*. Assume you have a process which at some point of time needs to wait for something to happen, such as a timer to fire. A task will simply block and wait until this happens. The scheduler will continue calling on it yet nothing will happen. The other tasks will continue working normally. An event-based system will finish the current event altogether and once the timer fires, a new event will be generated and the associated event handler will be called.

proto-threads   The Contiki operating system is a mix of both worlds. It is based on events but it also implements *proto-threads*, which are like tasks but with shared memory. The big brother of porto-threads are threads, which are also tasks with shared memory and thus no memory safety. Proto-threads are simpler, work on top of an event-based system, and are purely optional. Thus, the programmer has the luxury of implementing threads, but needs to take care of the shared memory to ensure that no memory gets corrupted. This is not that complex when you use different variables for different proto-threads. The different proto-threads communicate between each other with events, which each of them can easily generate and listen to.

2.4 SIMULATORS

The book's introduction mentioned the possibility of using a simulator instead of real nodes. A simulator is a software system, running on a normal computer, which mimics the behavior of some other system and its interactions. For example, a sensor network simulator mimics the behavior of sensor nodes and their communication with each other. Figure 2.9 presents a screenshot of the Cooja simulator, which is the simulator of the Contiki operating system.

Cooja will be very useful to you as it offers the possibility to test your applications and algorithms before going to the real sensor node.

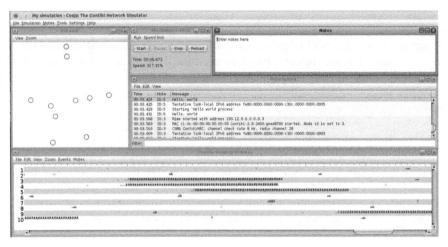

FIGURE 2.9   A screenshot of the Cooja simulator for Contiki based sensor nodes.

It takes the Contiki code that you have written and simulates its execution on virtual sensor nodes. The number of virtual nodes is not really limited, but high numbers cause longer simulation runs and need more resources. Interactions between the nodes are also simulated, e.g., packets are sent from one node to another or sensor readings become available. Besides the software which you provide for testing, a simulator also needs simulation models to work efficiently and to be of some use to the programmer.

Wireless propagation model is modelling how packets are transferred through the wireless communication medium. It is especially important to grasp the errors of the wireless medium, such as packet loss and packet corruption. The simplest model is the so called *unit disk graph model*. It says that each sensor node can send a packet to another sensor node, if the receiver is inside a certain circular area around the sender. The reception is always errorless. This model is typically too simple and simulation runs with this model lead to the impression that everything is working well and no packet corruptions occur or that between two nodes with a constant distance the sending of packets always work. This is not that simple, as we will see in Chapter 3. **unit disk graph model**

Mobility model is needed to define how sensor nodes are moving around the environment. This model is very useful when also your normal sensor nodes are moving, for example when they are attached to bikes or buses. There are some very simple models, such as the *random waypoint*. It always selects a new random point somewhere in the virtual environment and lets the sensor "drive" there with some constant predefined speed, then selects a new one, etc. There are also some more sophisticated ones, which for example follow streets on the map, take into consideration the habits of people, etc. **random waypoint**

The energy expenditure model is important to understand how much energy a sensor node would need when running specific software. As discussed in Section 2.2, the software controls the energy expenditure of the system.

The traffic model dictates how many events have occurred in the environment. Sensor networks are typically sensing something, e.g., temperature, rain volume, etc. It is important to understand when events worth sending are generated in a real environment and to also use similar values for simulation. For example, if the application is trimmed to report only temperature data below 15 and over 25 degree Celsius, the traffic model needs to realistically mimic how often and in which sequence such events realistically occur.

> A simulation is a set of different models, which together attempt to realistically mimic the sensor nodes' behavior and interactions. The more sophisticated the models used are, the less difference there is between simulation and real-world behavior of the implemented system. Very simplistic simulations are easy to program and understand but are not helpful when the system is transferred later to reality.

Simulation models are out of scope for this book, even if they are highly interesting topics by themselves. The interested reader can turn to Wehrle *et al.*'s book [6].

2.5 COMMUNICATION STACK

The software of a sensor node is typically organized into smaller pieces called modules. Instead of implementing everything into one mixed application, often called spaghetti code, programmers organize their code into modules, which can be easily (or at least more easily) reused with other applications. The communication stack is such an organization, where different tasks of the inter-node communication are organized into pieces.

The most widely used communication stack is the OSI model, which underlies the Internet. It consists of seven layers. The lowest one is the physical layer, where physical communication takes place (e.g., electromagnetic waves for wireless communications). The highest layer is the application such as an email client or a browser. In between, there are five other layers handling things such as who is talking when and how to route packets from distant nodes. More information about the OSI model and all its layers can be found in Tanenbaum's textbook [5].

2.5.1 Sensor Network Communication Stack

The OSI model is slightly too complex for sensor networks. Thus, we use the so called simplified OSI model presented in Figure 2.10. It consists of the following layers, which are connected to each other:

- **The application** layer is the top-most layer and is the actual application of the sensor node. For example, it can regularly sense the temperature and humidity

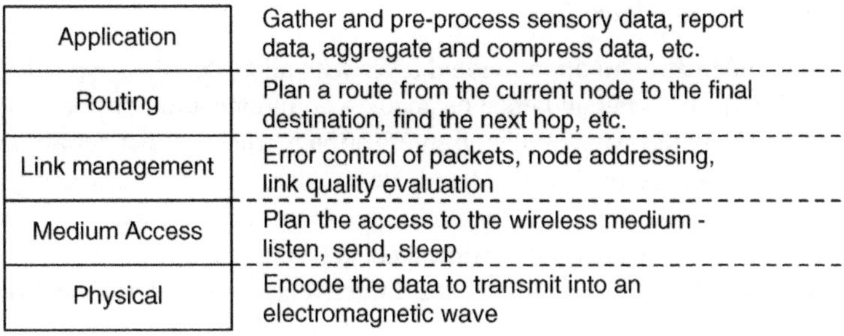

Application	Gather and pre-process sensory data, report data, aggregate and compress data, etc.
Routing	Plan a route from the current node to the final destination, find the next hop, etc.
Link management	Error control of packets, node addressing, link quality evaluation
Medium Access	Plan the access to the wireless medium - listen, send, sleep
Physical	Encode the data to transmit into an electromagnetic wave

FIGURE 2.10   The simplified OSI model as typically used for sensor network applications.

of an environment, pre-process the data (e.g., compare it to some thresholds), compress the data, and send it to the network's sink. Note that at this communication level, no information exists about where the sink is or where the node itself is.

- **Routing** is the layer which plans the communication in a broader scope. It knows how to reach the sink, even if it is several hops away (every node used to forward a packet to the sink is known as one hop). The routing layer abstracts the network as a communication graph, where paths between individual nodes exist. Chapter 5 covers routing in more detail.

- **Link management** covers the details about node-to-node communication. For example, it evaluates the communication quality between individual nodes; it corrects errors resulting from communication; and it takes care of giving addresses to nodes. Chapter 4 covers link management in greater detail.

- **Medium access (MAC)** is the layer where the node manages its access to the wireless medium. Since the wireless medium (the air) is a shared medium, it is important to decide exactly when to listen, send, sleep, etc. Chapter 3 discusses MAC in more detail.

- **Physical** layer is the layer where the bits and bytes of individual packets get transformed into an electromagnetic wave that is physically transferred from one node to another. Chapter 3 looks at some of the basic principles of wireless communications.

Each of these layers is assumed to work independently from the others and to communicate with them via well-defined interfaces. The main goal of this modularization is to have different implementations of individual layers, which can be combined freely in the context of different environments or applications. Of course, reality is never perfect and often one implementation depends on the implementation of something else in a different layer. Also quite often, two or even more layers are implemented together in one protocol, e.g., routing and link management. However, the differentiation between these layers in this book will greatly help you better understand and master the challenges of sensor networks.

2.5.2   Protocols and Algorithms

The exact implementation of an individual layer with all its details is typically called a protocol. Similar to a king audience protocol, a communication protocol specifies exactly and precisely the steps of all communication partners so that communication is successful. Algorithms, on the other hand, are smaller components, which typically compute something but do not specify communication steps. For example, a routing protocol specifies the steps needed to route a packet from one node to another remote one. An internal algorithm evaluates the effectiveness of different options and selects one.

Upcoming chapters introduce many different algorithms and protocols for all the mentioned communication stack layers. The next chapter starts from the lowest layer, the physical one.

ANATOMY OF A SENSOR NODE: SUMMARY

A sensor node consists mostly of a microcontroller, radio transceiver, some sensors such as temperature or humidity, a serial connection for debugging and programming, some LEDs for debugging, and sometimes an external flash memory to store sensor readings.

Each of these components need energy to operate and the most power-hungry ones are the radio and flash memory. These need to be used very sparingly and to spend most of their time in sleeping mode to save energy.

Toward the end of the battery lifetime, individual components **start degrading** but do not necessarily stop working. Instead, they can only support some of their functionality (the radio can send, but not receive) or deliver faulty results (the sensors).

Sensor node operating systems are simpler than modern operating systems such as Linux. They only support **static memory management** or some very limited version of dynamic memory. They also support **interrupts** for managing external components such as sensors. Contiki is a mixture of modern concepts such as thread-based programming with an event-based kernel.

FURTHER READING

Cormen *et al.*'s book [1] offers an excellent introduction to the theory of algorithms and how to evaluate their performance. Similarly, Tanenbaum and Wetherall's book [5] is an excellent introduction to computer networks and their most important concepts and examples.

Mottola and Picco's survey [2] gives a very comprehensive overview of existing programming paradigms for sensor networks, including languages, operating systems, and abstractions. An interesting example of a blocking implementation for operation management is the Scatterwerb system from the Free University in Berlin, described in Schiller *et al.* [4]. The sensor node lifetime study, which was discussed in Section 2.2, is presented in greater detail in Nguyen *et al.* [3].

A good and in-depth overview of sensor network simulation and models is given in the book of Wehrle *et al.* [6].

[1] T. H. Cormen, C. E. Leiserson, R. L. Rivest, and C. Stein. *Introduction to Algorithms*. MIT Press, 2009.

[2] Luca Mottola and Gian Pietro Picco. Programming wireless sensor networks: Fundamental concepts and state of the art. *ACM Computing Surveys*, **43**(3):19:1–19:51, April 2011. ISSN 0360-0300.

[3] H. A. Nguyen, A. Förster, D. Puccinelli, and S. Giordano. An Experimental Study of Sensor Node Lifetime. In *Proceedings of the 7th IEEE International Workshop on Sensor Networks and Systems for Pervasive Computing (PerSens)*, Seattle, WA, USA, 2011.

[4] J. Schiller, A. Liers, and H. Ritter. ScatterWeb: A wireless sensornet plattform for research and teaching. *Computer Communications*, **28**(13):1545–1551, August 2005.

[5] A. S. Tanenbaum and D. J. Wetherall. *Computer Networks*. Prentice Hall, 2010.

[6] K. Wehrle, M Günes, and J. Gross, editors. *Modeling and Simulation for Network Simulation*. Springer, 2010.

3

RADIO COMMUNICATIONS

The heart of wireless sensor networks is their ability to communicate wirelessly with each other. The most broadly used interface is the radio transceiver operating in one of the free bandwidths, which are reserved worldwide for research and medical applications. This chapter explores how a radio transceiver works and how this work impacts other communications and applications in WSNs. The focus here is on the basics of the communication stack's physical and medium access layers, providing the minimum yet absolutely necessary knowledge for handling wireless communications in WSNs.

If you are broadly familiar with radio communications or you have little interest in going into these details, you may consider jumping to Section 3.3, where medium access protocols are discussed, or you might go directly to this chapter's summary. The summary, however, is highly recommended because some peculiarities of radio communications greatly impact WSNs in their work.

3.1 RADIO WAVES AND MODULATION/DEMODULATION

Radio waves are just normal electromagnetic waves. Their name refers to their frequency range in the electromagnetic spectrum (Figure 3.1). Any electromagnetic wave has three parameters (Figure 3.2) and follows the equation:

$$s(t) = A(t)\sin(2\pi f(t)t + \phi(t)) \qquad (3.1)$$

Introduction To Wireless Sensor Networks, First Edition. Anna Förster.
© 2016 The Institute of Electrical and Electronics Engineers, Inc. Published 2016 by John Wiley & Sons, Inc.

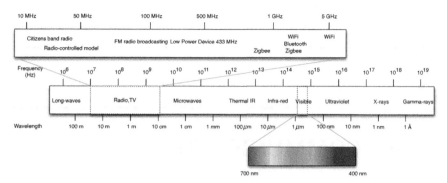

FIGURE 3.1   The full electromagnetic spectrum.

The term radio only indicates the approximate frequency of the wave or part of the electromagnetic spectrum where it resides.

A natural wave in itself does not carry any information. In order to encode some information into it for data communication, you must change the parameters of the radio wave in a well-defined way so these changes can be detected at the receiver side and the same information can be decoded. This is similar to writing text with letters and sending it via the mail. At both sender and receiver sides, you need the same encoding system, e.g., the English language with Latin letters. Thus, the encoded text can be decoded in the same way at the receiver side. This process is called modulation/demodulation.

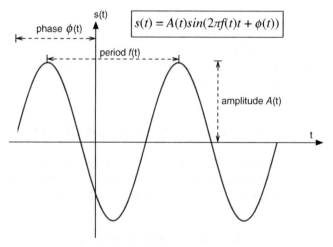

FIGURE 3.2   The three main parameters of an electromagnetic wave: amplitude A(t), frequency or period f(t) and displacement or phase ϕ(t).

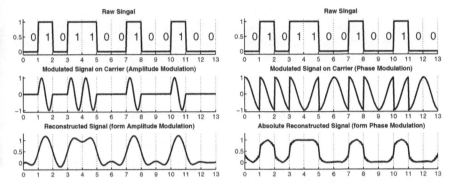

FIGURE 3.3   By changing the parameters of the radio signal (wave) we can encode information into it. This process of encoding and decoding is called signal modulation/demodulation. In the figure we can see how zeros and ones can be encoded into a radio wave, by using two of three possible parameters – amplitude on the left and phase on the right.

Definition 3.1. *Signal modulation/demodulation.* *This is the process of changing radio wave parameters in a well-defined way to encode/decode information into/from the wave.*

Any of the radio wave's three parameters, or combinations of them, can be used to modulate the signal (Figure 3.3):

- **Amplitude** A(t). This parameter gives how high the wave is. To encode information, you can change the amplitude from very small (encoding a 0) to very high (encoding a 1).
- **Frequency or period** f(t). This parameter dictates how often the wave form is repeated over time. The frequency of the signal can be changed to indicate different codes.
- **Displacement or phase** ϕ(t). This parameter identifies the displacement of the wave in respect to the beginning of the axes. You can displace the wave to indicate change of codes.

A *modulation code*, or key, is the symbol that you are encoding onto the modulation code wave. For example, if you decide to use a very high amplitude to encode a 1 and a very low amplitude to encode a 0, you have two codes or keys. Of course, more than two keys or (ones and zeros) can be encoded into the signal by using more than two levels of amplitude or more than two different frequencies. Combinations of different modulation codes also lead to many additional modulation codes.

The process of modulation and demodulation is rather simple to understand and use. In fact, some kind of wave modulation is used for all wireless communications.

modulation
code

But knowing from everyday life the problems connected to wireless transmissions, why do they sometimes not work as expected? Problems arise from the wave propagation properties through your environment, or in other words, what remains from the wave after it travels some distance through the environment (air, water, free space, etc.).

3.2   PROPERTIES OF WIRELESS COMMUNICATIONS

wave propagation While traveling through the environment (we talk about *wave propagation*), the electromagnetic wave experiences multiple distortions. These are mainly due to the following processes, which are shown in Figure 3.4.

- **Attenuation.** This process spreads the energy of the wave to larger space. It is similar to a balloon, which is a dark red color before filling it with air, but then becomes almost transparent once filled. Thus, with growing distance from the sender, the wave becomes less and less powerful and harder to detect (Figure 3.4a).

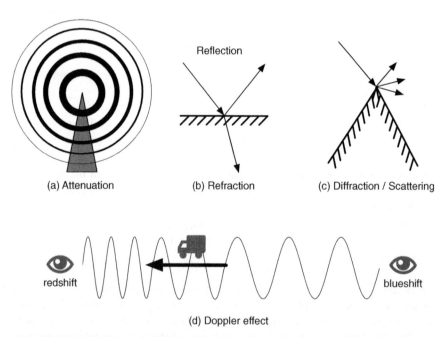

(a) Attenuation   (b) Refraction   (c) Diffraction / Scattering

redshift   blueshift

(d) Doppler effect

FIGURE 3.4   Different physical processes that lead to path loss in signal propagation. (a) Attenuation spreads the power of the signal over a larger space. (b) Reflection and refraction changes the trajectory of the wave and generates secondary waves. (c) Diffraction generates secondary waves on sharp edges. (d) Scattering occurs when the signal hits a rough surface and deviates from its original trajectory. (e) Finally, the Doppler effect shifts the frequency of the signal depending on the direction of movement related to the receiver.

What does this mean for the modulation? Imagine you have used an amplitude modulation method and fixed an amplitude X for encoding a zero and Y for one. You will use these amplitude levels for encoding information at the sender. However, at the receiver, you will never exactly detect these levels. This complicates the process of demodulation because it requires you not to look for particular levels of amplitude, but for differences between them.

- **Reflection/refraction.** This process changes the direction of the wave when it meets a surface. Part of the wave gets reflected and travels a new trajectory, another part of the wave gets refracted into the material and changes its properties. Both processes create new, secondary waves, which also reach the receiver at some point in time, slightly after the primary wave. This is both a blessing and a curse—very weak signals can be better detected when the primary and the secondary signals overlap. At the same time, it is hard to understand what a secondary wave versus a primary wave is (Figure 3.4b).

- **Diffraction/Scattering.** Sharp edges and uneven surfaces in the environment can break the wave into several secondary waves with the same consequences as above (shown in Figure 3.4c).

- **Doppler effect.** In general, the frequency of the signal changes with its relative velocity to the receiver. The Doppler effect is well known for its impact on the police siren, which sounds different to the observer depending on whether the police car is approaching or moving away. The same happens with the radio waves when their frequencies get shifted in one or the other direction which results in a loss of center. For wireless communications, this means that you can hardly differentiate between various frequency codes (Figure 3.4d).

All of these processes lead to what researchers summarize as path loss.

Definition 3.2. *Path loss is the reduction in power density of an electromagnetic wave as it propagates through space.*

Path loss is central to wireless communications, as it allows you to predict the quality of the transmission and/or to design wireless links. In the remainder of this section we will explore how path loss behaves in reality and impacts wireless communications.

3.2.1 Interference and Noise

So far this chapter has only considered wireless transmission problems resulting from a single sender and its environment. However, typically there are several simultaneous senders and other even non-manmade sources of disturbance. The latter is also referred to as noise.

> **Definition 3.3.** *Electromagnetic noise is the unwanted fluctuation of a signal or energy from natural sources such as the sun. It is generally distinguished from interference or from systematic alteration of the signal such as in the Doppler Effect. It is typically measured with signal-to-noise ratio (SNR) to identify the strength of the useful signal compared to the overall environmental noise.*

On the other hand, the corruption of a signal through other active senders is called interference.

> **Definition 3.4.** *Electromagnetic interference is the disturbance of an electro-maggnetic single due to an external source. It is typically measured with the signal-to-interference ratio (SIR) or with signal-to-noise plus interference ratio (SNIR).*

The last metric, SNIR, is used more often, as it combines both noise and interference. These two are not easily separable when making measurements and combining them is useful and simple.

In general, interference is something you try to avoid in wireless communications, as it increases the path loss and leads to completely lost signals. In terms of wireless sensor networks, do keep in mind that interference is caused generally by other electromagnetic waves, not only by other sensor networks. Thus, microwaves, Wi-Fi access points, Bluetooth, etc., are all sources of interference. Interference also leads to two famous problems in wireless communications: the hidden terminal problem and exposed terminal problem.

3.2.2   Hidden Terminal Problem

The hidden terminal problem is best explained with a diagram (Figure 3.5). As you can see, there are four nodes present. Node A starts sending packet X to node B. Node C is situated outside of the *transmission range* of node A, so it does not know anything about the ongoing transmission of packet X. Transmission range is generally referred to as a semi-circular area around a sender where an ongoing transmission can be detected.

transmission range (margin note)

Remark 3.1. *Note that the transmission range is often illustrated as a perfectly circular area around a sender, but is rarely circular. Typically, it resembles more of a toothed semi-circular area, something highly irregular in different directions. This irregularity comes from errors in the hardware, varying power at the radio transceiver, and irregular environmental patterns around the sender. More information about this can be found in [10].*

Since node C does not know anything about the ongoing transmission between A and B, it starts to transmit a packet to node D. This causes interference at node B, which corrupts packet X. However, the transmission between C and D is successful.

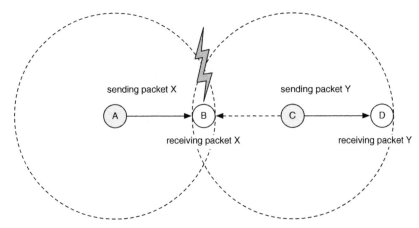

FIGURE 3.5   The hidden terminal problem in wireless communications. Node A tries to send packet X to node B. At the same time, node C decides to send to node D, because it does not hear anything from node A (out of transmission range). The interference of both packets at node B results in packet loss. However, the packet at node D is received.

The hidden terminal problem is a major challenge to overcome in wireless communications. Later in this chapter in Section 3.3.3, you will see how to resolve it and to let neighboring nodes know about ongoing transmissions.

Remark 3.2. *Interference is sometimes a difficult problem. At times, nodes are not able to receive an error-free packet from some particular node but its transmissions still cause interference. This greater area is often referred to as* interference range. *Because of this, the hidden terminal problem will never be completely resolved.*

interference
range

3.2.3   Exposed Terminal Problem

The exposed terminal problem is the opposite of the hidden terminal problem. Figure 3.6 illustrates the case, when a node is prevented to send a packet because of another ongoing transmission. This time, node B is sending a packet to node A. At the same time, node C has a packet to send to node D. Looking at the topology of the network and transmission ranges, you can see that both transmissions are simultaneously possible. However, node C cannot send a packet because it hears the ongoing transmission of node B and does not know where node A is. Thus, it needs to wait, even if the transmission is possible.

3.3   MEDIUM ACCESS PROTOCOLS

The role of Medium Access Protocols is to regulate the access of the sensor nodes to the shared wireless medium, this is, to the "air". However, we will first define some important metrics, which will help you identify how well a medium access protocol (*MAC protocol*) is performing.

MAC
protocol

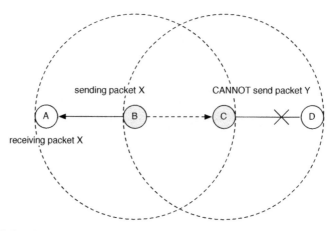

FIGURE 3.6   The exposed terminal problem in wireless communications. Node B sends a packet to node A. At the same time, node C wants to send a packet to node D, and as the transmission ranges suggest, both transmissions are possible simultaneously. However, node C is prevented from sending, because it hears the transmission of B but does not know where A is.

> **Definition 3.5.** *Throughput is defined as the number of bits or bytes successfully transmitted per time unit. Typically, you will use bits per second. The throughput can be defined over the medium itself (cable or wireless), over a link (between two communicating nodes), or only for a single node.*

Note that the throughput can be different in both directions—for sending and receiving. The throughput shows how much data you can push through a given medium whether cable, wireless, or even a sensor node. In terms of channel throughput, your interest should be in how many bits or bytes you can send over this channel so that all of them successfully arrive at the other side. When looking at the throughput of a sensor node, your interest should be in how many packets you can process and send out without issue. A MAC protocol typically attempts to maximize the throughput both at individual nodes and for the wireless medium in general. Furthermore, it **fairness** tries to ensure some sort of *fairness*. This means that each node should have a fair chance to send out its packets.

Another rather useful metric is the delay.

> **Definition 3.6.** *Delay is the amount of time between sending a packet and receiving the packet. The delay can be defined between any two communicating components – internal hardware or multi-hop end-to-end communications.*

TABLE 3.1   Nominal power consumption of Z1 nodes.

Component	Mode	Current draw
Microcontroller	Active	1.8 mA
	Sleep	5.1 μA
RF Transceiver	Receive	19.7 mA
	Transmit (at 0 dBm)	17.4 mA
	Sleep	0.01 mA

Data from www.memsic.com and www.ti.com

Delays can be very short when measured internally between two hardware components and extremely large, e.g., when measured between terrestrial and satellite nodes. The delay provides you with information on how much time passes between creating a packet and its real arrival at the destination. A MAC protocol tries to minimize this time for all involved parties.

The main goal of medium access protocols is to *prevent* interference and corrupted packets, while *maximizing* the throughput of the wireless medium and *minimizing* the energy spent.

3.3.1   Design Criteria for Medium Access Protocols

We are now ready to look at the design criteria for good MAC protocols. A MAC protocol has a hard problem to solve. It needs to maximize the throughput by minimizing the delay and energy spent. Additionally, for sensor networks, it needs to be able to handle switched off devices so nodes do not waste any valuable energy. As discussed in Chapter 2, energy is the greatest problem. Let us explore how much energy a typical sensor node uses for its rudimentary operations (Table 3.1).

Not surprisingly, the node spends much less energy in sleep mode than in active mode. However, what is more intriguing and less intuitive is that it spends more energy while receiving or even simply listening to the channel than while sending. This means you need to minimize the time in listening mode as much as possible. This also leads to the four main design criteria for medium access protocols.

- **Minimize collisions.** By avoiding packet collisions, the MAC protocol also avoids resending packets, which of course increases the throughput and decreases the energy spent. This task is rather difficult as the MAC protocol needs to orchestrate all nodes. Recall the hidden terminal and the exposed terminal problems from Sections 3.2.2 and 3.2.3.
- **Minimize overhearing.** Overhearing is when a node receives a packet, which was not destined for it. Sometimes this can be useful, e.g., to see what is happening around. However, there are smarter ways to control who is receiving the

packet and who is not (by using destination addresses in the header, for example). Thus, overhearing needs to be avoided because the overhearing node is supposed to trash packets which are not destined to it. Again, this is a hard task because the node needs to know when a packet is destined for it and when not and to go to sleep accordingly.

- **Minimize idle listening.** This problem is similar to overhearing in terms of energy waste because the node uses the same amount of energy in idle listening and receiving modes. It refers to the mode when the node is simply listening to the channel and nothing happens. This time needs to be minimized to save energy. Typically, going to sleep is not a problem but deciding when to wake up again is more challenging.

- **Minimize overhead.** Every packet and every bit, which does not carry application data (such as temperature or events), is considered an overhead. Even the destination and sender addresses are overhead. Each bit uses energy when sent or received and thus needs to be avoided, if not absolutely necessary. Completely avoiding the overhead is considered impossible with current technologies and is left to future generations to achieve (feel free to attack the problem).

How is it possible to solve so many problems at the same time? There is no universal cure and various protocols typically solve one or several of the problems and neglect or trade off others. New protocols are continuously published. However, there are several possibilities to approach medium access and some protocol families have also formed over the years. Next, important and widely used protocols are discussed and later on in Section 3.3.7, you will discover what else is out there.

3.3.2 Time Division Multiple Access

The easiest way to organize communications in a network is by time. This is also called time division multiple access (TDMA). The basic principle is similar to "divide and conquer": Divide the time available across the nodes and give them full control over their slots.

To achieve this, time is divided into rounds and rounds are divided into slots (Figure 3.7). Each node is then given sending control over one slot. The number of slots

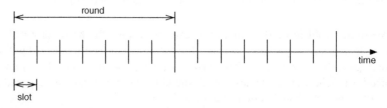

FIGURE 3.7 Rounds and slots in a TDMA schedule. The length of the slot depends on the technology used, clock precision of the sensor nodes, and on the expected length and number of the packets to be sent in one slot.

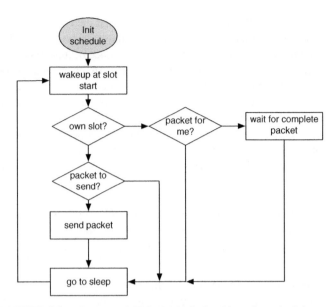

FIGURE 3.8   The general TDMA MAC algorithm after schedule setup.

depends on the number of nodes in the network. The length of one slot and thus the length of a round depends on the technology used and on the expected length of the packets. It also depends on the precision of the sensors' internal clocks.

Figure 3.8 presents the generalized TDMA algorithm. However, this is the algorithm for how TDMA works *after* the schedule has been set up. There are two possible ways to set up the schedule: centralized and distributed.

Centralized TDMA. In centralised TDMA, the schedule is calculated offline and provided to the sensor nodes at startup. If nothing is known about the topology or the individual links between the nodes, the only option is to provide exactly N slots for N nodes and to give them control over the slots with their corresponding ID. Of course, if information about the topology, possible links and interferences is available, the task becomes more complex in computation, but the result can be more efficient. For example, non-interfering nodes can use the same slot and thus minimize the delay in the system.

The main disadvantage of centralized TDMA is its rigidness. Especially if the schedule takes into account the topology or the individual links between nodes, any slightest change in those will result in a complete mess in the schedule and harsh interference. On the other side, if the network is small and traffic low, centralized TDMA with N slots for N nodes is probably the simplest way to avoid interference and to allow for sleep.

Distributed TDMA. In the distributed variant of TDMA, the nodes attempt to find a good schedule by cooperation. Typically, the system starts with a predefined number and length of slots. The nodes start in the CSMA communication style, which means that they compete for the channel (that style is explored next). They first exchange neighbor information in terms of link quality. Next, they compete for the slots by trying to reserve them then release them again if interference occurs. When the system stabilizes, the network is ready to enter the final TDMA phase and to start working collision-free. Distributed TDMA is much more flexible and can also take into account the nodes' individual traffic needs. However, the initialization phase is quite complex and there is no guarantee that it will successfully converge. At the same time, once the schedule is found and no changes in the topology occur, communication is very efficient and fully collision-free. Even if changes occur, the protocol is able to detect them and to restart the initialization phase, if needed. An example of such a protocol is TRAMA [7].

Remark 3.3. *Note that TDMA is an approach, which is implemented by many different protocols. TRAMA is a specific protocol implementing further details such as slot duration, competition phase, various parameters, and so on.*

Discussion. What are the advantages of this time-division scheme? Every node knows exactly when it is allowed to send data (in its own slot only). Thus, there will be no more collisions in the system. Plus, every node can first easily send a list of intended recipients before sending the data, thus enabling the others to go to sleep, if they are not on the list. This is similar to entering a crowded office, everybody looking up and you gesturing to the person with whom you would like to speak. Everybody else goes back to their work and the intended person listens to what you have to say. Another plus is that this system is perfectly fair in terms of sending time available for each node. But is this efficient?

First, sometimes nodes will not have anything to send so their slot is wasted. Second, and even more importantly, some nodes inherently have more traffic. Think about a simple scenario like the one in Figure 3.9. All of the gathered data needs to be transferred to the sink. This cannot be done directly as the sink is too far away from some nodes. This process is called routing and Chapter 5 explores it in greater detail. For now, just assume that the packets need to be transferred along the tree-like structure

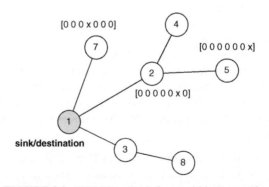

FIGURE 3.9 TDMA example in a multi hop scenario.

as shown in the figure. Thus, node 2 is now responsible to send its own packet to the sink (one hop) and also to resend the packets of nodes 4, 5, and 6. These nodes only need to send their own packets. Consequently, fairness is if node 2 had at least 3 slots and nodes 4, 5, and 6 only one slot each.

Another problem with TDMA is the slot sequence. In Figure 3.9, the slot assignments of the individual nodes are provided in squared brackets next to the nodes. If node 5 sends its packet to node 2 in slot 7 (the last of the round), but node 2 has control over slot 6, the packet has to wait almost a full round of slots before it can proceed. Thus, depending on which slots belong to which node and on the sequence of transmissions, some packets will go fast through the network whereas some will be extensively delayed.

3.3.3   Carrier Sense Multiple Access

Carrier sense multiple access (CSMA) refers to a simple, but powerful protocol, which works on the principle of "listen before talk". This means the sender first listens on the shared channel and if it is free, it tries to send. Two main variants of CSMA exist: CSMA with collision detection (*CSMA-CD*) and CSMA with collision avoidance (*CSMA-CA*). CSMA-CD tries to detect a collision, and if it happens to resend the packet. The second one tries to avoid the collision in the first place. This discussion focuses on CSMA-CA since it is more often used and performs better.

CSMA-CD

CSMA-CA

Figure 3.10 shows the flow diagram of the CSMA-CA protocol.

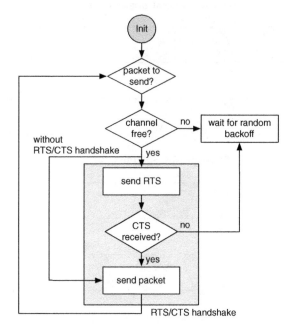

FIGURE 3.10   Flow diagram of the general CSMA with collision avoidance (CSMA-CA) protocol.

All nodes are assumed to be always in idle listening mode. When a node has a packet to send, it first listens for traffic on the channel. If the channel is free, it initially sends a special message, called ready-to-send (RTS). The receiver sends a short **RTS-CTS** answer called clear-to-send (CTS). This procedure is also called *RTS-CTS handshake* **handshake** because it resembles the handshake when two people meet and before they start talking to each other. When the sender receives the CTS packet, it is ready to send the application data (the real packet) and can optionally listen for an acknowledgment or ACK for short. If the channel is busy, the sender waits for some random amount of **random** time (*random backoff*), until the channel becomes free again. **backoff**

How does this handshake help avoid collisions? Figure 3.11 illustrates some examples. In the first scenario, (A), everything works as intended. The RTS/CTS messages

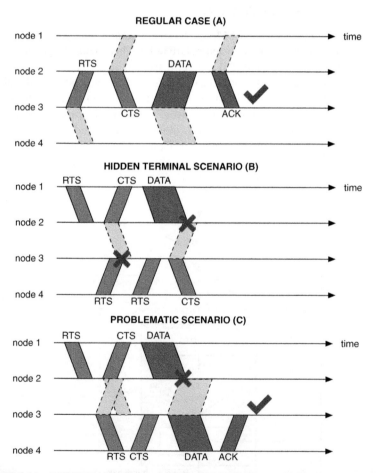

FIGURE 3.11 RTS/CTS handshake of CSMA-CA in various scenarios. (A) shows the normal, intended operation. (B) shows the hidden terminal scenario. (C) shows another problematic case.

successfully shut down the nodes around (nodes 1 and 4) to not disturb the communication between nodes 2 and 3. This is how the collision avoidance is supposed to work—by shutting down nodes around the sender and the receiver. However, this does not always work. In scenario (B), you see how node 4 attempts to send a RTS at the same time as node 2 sends its CTS to node 1. These two packets collide only for node 3, so it does not get anything and, in turn, does not answer to node 4. Node 4 only knows that something went wrong. But what? It backoffs randomly and retries after some time. The RTS is now successfully received at node 3 and its answers with a CTS, which interferes at node 2 with the data from node 1 and interrupts their communication. This shows a clear example of the hidden terminal problem.

Focusing on scenario B, node 3 would have been successfully shut down, if it would have received the CTS from node 2. However, it collided with the RTS from node 4, which prevented node 3 from shutting down. Thus, CSMA-CA does not really solve the problem of the hidden terminal, but only ameliorates it.

Scenario C is very similar. Here, the RTS and CTS packets from nodes 2 and 3 collide, and node 3 is not properly shut down. Communication continues between nodes 1 and 2 and between 3 and 4, until the data of both collide.

Now, the last question to answer is, what to do if the channel is busy and you cannot start immediately sending RTS/CTS? There are several options.

- **1-persistent CSMA.** If the channel is busy, the sender continues sensing the channel until it becomes idle again and sends them immediately. If a collision occurs, it backoffs for a random amount of time and re-attempts. The problem is that if several senders are waiting for the channel to become free, they all attempt at exactly the same time and only the random backoff solves (hopefully) the problem. Thus, this version is in fact CSMA with collision detection rather than collision avoidance. It is used for example in Ethernet.
- **Non-persistent CSMA.** This version is the opposite of 1-persistent and much less aggressive. If the channel is busy, it backoffs for a random amount of time and then retries, instead of continuously sensing the channel. This leads to much less collisions, but also to larger delays.
- **P-persistent CSMA.** This is a tradeoff between 1-persistent and non-persistent CSMA. If the channel is busy, the sender continuously senses it until it becomes idle again. Here, however, it only sends the packet with a probability P. If it chooses not to send, it waits for some small amount of time and then repeats the procedure. This version leads to better use of the channel and at the same time has lower delays than non-persistent. It is used for example in Wi-Fi, the lower layer of *Zigbee, IEEE 802.15.4*. **Zigbee, IEEE 802.15.4.**
- **O-persistent** CSMA. This version is time scheduled. A central controller in the network assigns a fixed transmission order to the nodes and they wait for their turn for transmitting, thus it is a TDMA approach.

Overall, CSMA is a very simple and understandable protocol and works rather well in most situations. However, its main disadvantage in the context of sensor networks

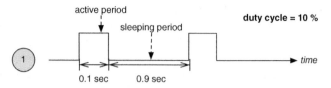

FIGURE 3.12   Active and sleep periods of a sensor node, also called duty cycle.

is its energy consumption. It never puts the nodes to sleep and will exhaust the energy on a sensor node very quickly (typically a couple of hours).

3.3.4   Sensor MAC

Sensor MAC (S-MAC) has been developed especially for sleep-enabled sensor networks. It allows nodes to go to sleep and to perform their communication only during their active or awake cycles. This is illustrated in Figure 3.12, which is the preferred way of functioning for sensor nodes because it saves a lot of energy. The relationship between active and sleeping time is referred to as duty cycle.

Definition 3.7. *Duty cycle is the relation between the length of the active and sleeping cycles of a sensor node and is measured in percent. It is defined as:*

$$duty\_cycle = \frac{time\_active}{period}$$

In Figure 3.12, the duty cycle would be exactly 10%, as $\frac{0.1}{1.0} = 0.1 = 10\%$.

The main question is how to synchronize the nodes so that their sleeping cycles fall together and all are awake when somebody is sending. Recall that nodes cannot receive anything when sleeping. The system is deployed with a predefined sleeping interval, e.g., 1 second. Every node boots individually and listens for some time in a CSMA style to the channel. If it does not receive anything, it just picks up a schedule (a timestamp to go to sleep), sends it out and goes back to sleep for the predefined period of time. Next time it wakes up, it will first send its own schedule out again, then packets, and afterward go back to sleep again. If the newly booted node receives a schedule from another node, it joins this schedule by going to sleep at exactly the same time. From now on, both nodes are synchronized and will always wake up together. This scenario is depicted in Figure 3.13.

What happens if a node receives two different schedules from nodes that cannot hear each other? It needs to follow both schedules and wake up twice as often as the other nodes. Such nodes are often referred to also as **bridges** between

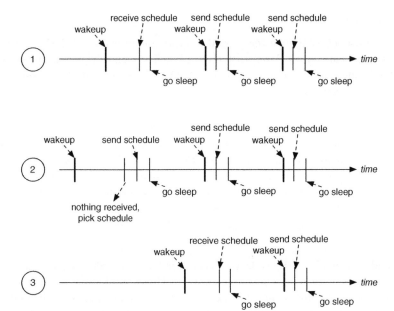

FIGURE 3.13   Sensor MAC general scenario.

synchronized islands. After synchronization, each node can start following its own **synchronized** schedule(s). It wakes up, sends out its schedule again for re-synchronization, and **islands** then uses an RTS/CTS handshake to send and receive packets.

Variations and Discussion.   S-MAC was the first protocol to allow sensor nodes to sleep for extensive amounts of time. However, it also suffers from some inherent problems. The first one refers to the internal clock synchronization of the nodes. Internal clocks are rarely reliable, especially because the used hardware is cheap and simple. This leads to clock skews and what a particular sensor node assumes to be 1 second, might be 2 seconds for another sensor, but is in reality 0.8 seconds (an exaggerated example). Experiments have shown that if no continuous clock synchronization is used, sensor nodes drift away completely from each other in approximately 30 to 60 minutes. Thus, time synchronization is required for S-MAC to work properly, which Chapter 7 discusses in detail.

Another problem lies in the protocol design itself. In their active periods, sensor nodes are expected to follow the CSMA/CA protocol. Thus, they stay idle even if there is no communication going on. Timeout-MAC (*T-MAC*) tackles this prob- **T-MAC** lem by letting the nodes timeout, when nothing happens for some time during their active periods. They will wake up later again at their normal wakeup time. Another problem for S-MAC are the bridges between the synchronized islands. Those nodes need to wake up twice or even more often than other nodes and exhaust their energy reserves much faster. One possibility is to reschedule one of the islands but the protocol becomes complex and fragile.

3.3.5 Berkeley MAC

B-MAC Berkeley MAC (*B-MAC*) was developed to tackle the problems of S-MAC. Some-
LPL times it is also called Low Power Listening (*LPL*). The main design requirement was
to simplify the protocol and to get independent from synchronization issues. The
solution was long preambles:

Definition 3.8. *A preamble is a special communication message of varying
lengths, which does not carry any application data or other payloads. Instead, it
signals to neighbors that a real message is waiting for transmission. The preamble can carry sender, receiver, and packet size information or other administrative
data to simplify the communication process.*

The idea is simple: If the node has data to send, it checks the channel and if it is
clear, it starts sending a long preamble message. This is similar to CTS but it is long
enough to wake up other nodes. For example, if the sleeping period of the nodes is
1 second, the preamble needs to be sent for at least 1 second to make sure that each
neighbor has a chance to wake up. When waking up, the nodes sense the channel. If
there is a preamble on it, they do not go back to sleep, but wait for the real packet to
arrive. Once the preamble finishes, the sender can send directly the real data without
further information (Figure 3.14).

However, B-MAC suffers from high delay of transmission and energy waste to
send and receive long preambles. That said, the delay is usually not problematic for
most sensor network applications, as they are rarely real-time dependant. But the
energy waste needs to be handled.

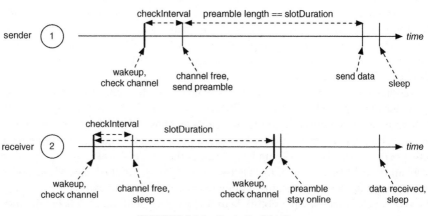

FIGURE 3.14 Berkeley MAC.

3.3.6   Optimizations of B-MAC

One possibility to tackle the problem of long preambles is to keep the sleeping period smaller. On one side, this is good to reduce the size of the preambles and also the idling time of the receivers. However, it increases the idling time when the node wakes up and there is nothing on the channel.

Another possibility is to optimize the work of B-MAC at least for some types of packets, namely when two nodes are communicating directly with each other and nobody else. This is also called unicast communication. In B-MAC, however, this is not taken into account and all nodes will receive the data packet. You can tackle this problem first from the side of the receiver.

If it hears a preamble with a destination different from its own ID, it can go immediately back to sleep and miss the packet. But what happens when the targeted receiver wakes up? It stays awake for a long time, until the end of the preamble—in the worst case, for a complete cycle. This can be solved by interrupting the preamble sender. The sender, instead of jamming the channel with the preamble, can send very short, identical preambles, and listen in between for an acknowledgment from the receiver that it is already awake. This will interrupt the preamble sequence and considerably lower the idling times and the total delay in the system. This variation of B-MAC is called *X-MAC* [2].   **X-MAC**

Other two variants are *BoX-MAC-1* and *BoX-MAC-2* [6]. They combine B-MAC   **BoX-MAC-1**
and X-MAC in a way to minimize idling and overhearing. Both BoX-MAC variants   **BoX-MAC-2**
mostly target implementation details such as at which level the receiver acknowledgment is sent to save energy and time. The algorithms and the principles are the same as in X-MAC and B-MAC. Most likely, this is why the variants have not been academically published.

Unfortunately, none of these optimizations work well for broadcast transmissions. One possibility is to wait for K receivers to wake up and then to send out the data packet. Another, more efficient way, is to skip the preamble all together and to send out the data packet many times. What happens at the receivers' side? Any receiver, when awake, directly receives the data packet and can go immediately back to sleep. This approach even positively impacts the energy budget of the sender, as it sends the data packet for the preamble interval and then immediately goes back to sleep, instead of sending the data packet again.

Yet another possibility is to inform the receiver when the data packet will be sent. This tackles the problem of broadcasting and unknown receivers, and also enables all potential receivers to sleep again until the real data transmission starts. MH-MAC is an example for such a protocol [1]. The only difference to X-MAC is that it attaches the expected time for data transmission to the preamble.

3.3.7   Other Protocols and Trends

There are so many MAC protocols developed for sensor networks, with most of them called XYZ-MAC, that people began talking about the MAC Alphabet Soup. This inspired a website (http://www.st.ewi.tudelft.nl/~koen/MACsoup/), which attempts to bring some order and offer a taxonomy. It was abandoned around 2010 but it still offers the best overview of MAC protocols available.

One often mentioned standard is IEEE 802.15.4. It defines the physical and the MAC layer for various standards. IEEE 802.15.4 is based on CSMA/CA and operates on various license-free channels, such as 868/915 MHz or 2450 MHz. It is interesting to note that many of the MAC protocols explored in this book actually work on top of the IEEE 802.15.4. For example, B-MAC simply switches on and off the radio and defines the length of the preamble period and uses IEEE 802.15.4 to actually transmit the packets and the preambles. Other protocols are lower and need access to the hardware directly such as A-MAC, described below. IEEE 802.15.4 is the basis for some industry-level communication standards, such as *Zigbee, ISA100 or Wire-lessHART*. These are preferably used in industrial settings, where a well-defined and already implemented protocol is more important than performance. However, those protocols are quite restrictive and waste a lot of energy for management issues, for example to maintain a communication structure on top of the wireless links. They also do not use any sleeping or duty cycling on the nodes, which makes them useful only for networks with constant energy supply.

Zigbee ISA100 Wire-lessHART

A-MAC [3] is considered the best algorithm ever invented and was presented by Dutta *et al.* in 2010 at one of the most prestigious conferences for sensor networking research, ACM SenSys, where it also won the best paper award. The main idea is to initiate the transmission on the receiver and not the sender. Thus, the sender just waits for the receivers to wake up and ping it. The receiver sends a short preamble, which is acknowledged by the sender at the hardware level exactly X nsec later. This hardware precision has the purpose of making several acks *non-destructive* or, in other words, to render the receiver able to receive them even if they collide. If the acks arrive with more than 500 nsecs difference, they destroy each other. A-MAC is able to efficiently organize the communication in a sensor network and to avoid delays, long preambles, and long idling periods. However, its implementation is very complex and requires tapping on the radio hardware. This is the main reason why it has remained rather an academic exercise.

non-destructive interference

Currently, BoX-MAC is considered the de-facto standard for real-world sensor networks. It is not as complex as A-MAC, is already implemented for almost all WSN operating systems, and is simple to understand and use. These are the winning properties for a good WSN protocol and not academic-level performance. At the hardware level, 802.15.4 is typically supported (part of the Zigbee standard). However, it does not offer any energy saving or sleeping options.

The trend generally goes towards generalized protocols, which are able to self-adapt to traffic conditions. There is still no consensus whether one should avoid broadcast transmissions altogether or should favor them. When selecting the right MAC protocol, it is typically advisable to carefully explore the communication requirements in terms of expected traffic (throughput, bursty or regular traffic, etc.) and types of transmissions (unicast or broadcast).

RADIO COMMUNICATIONS: SUMMARY

What did you learn in this chapter? Wireless communications are complex to capture and to manage. Their main properties are:

⇒ They are an error-prone process whose properties and quality fluctuate significantly with environment, distance, and time.
⇒ Interference between different nodes and other technologies greatly impact the quality of links.

Moreover, you need to actively control the access to the wireless medium from various sensor nodes to enable efficient, energy-preserving, and fast communication. A MAC protocol needs to enable the following properties:

⇒ Collision-free communication.
⇒ Minimal overhearing of packets not destined to the node.
⇒ Minimal idling when no packets are arriving.
⇒ Minimal overhead and energy for organizing the transmissions.
⇒ Minimal delay and maximum throughput of packets.

There are several general approaches you can take:

⇒ *Time Division Multiple Access* (TDMA) refers to a mechanism in which each node gets full control for some predefined amount of time (a slot). This is collision-free, but it suffers from large delays.
⇒ *Carrier Sense Multiple Access* (CSMA) refers to "first listen, then talk." While the delay is low, the energy expenditure is high (the nodes never sleep) and it is not collision-free.
⇒ *Duty cycling* is the preferred way of organizing the sleep and awake cycles of sensor nodes. Sensor MAC, Berkeley MAC, and BoX MAC all work with duty cycling and are able to save considerable amounts of energy.
⇒ *BoX MAC* is based on B-MAC, but offers optimized communications for both unicast and broadcast transmissions, and is currently the preferred MAC protocol for sensor nodes. It does not need synchronization, has low delay, and low energy expenditure.

QUESTIONS AND EXERCISES

3.1. Recall the general TDMA algorithm from Figure 3.8. Modify the algorithm so that several packets can be sent during one slot instead of only one. Discuss how this modification changes the energy expenditure of the algorithm.

3.2. Recall CSMA-CA from Section 3.3.3. Does it solve the exposed terminal scenario? Draw the scenario and discuss.

3.3. Recall T-MAC from Section 3.3.4. It tackles the problem of idling nodes when there is no communication going on. However, what happens if the active period of the nodes is long enough to accommodate more than one transmission? What happens to the second packet to be sent? Draw the time diagram and discuss.

3.4. Recall B-MAC from Section 3.3.5. The section discussed that it is independent from time synchronization. Clock problems can be differentiated into skew and precision. Discuss how these two properties, precision and skew, affect B-MAC. How much do they interfere with its work?

3.5. Recall B-MAC from Section 3.3.5. In its variations and optimizations, you learned the possibility to keep the sleeping period short to allow for shorter preambles and idling periods. Consider a simple scenario, where 2 nodes communicate to each other. Node A has a packet to send to node B every second, and the transmission of the packet itself takes 0.1 seconds. Node B only receives the packets from node A and is itself silent. Assuming the energy consumption rates from Table 3.1, calculate the energy spent by both nodes for a total time of 1 minute in two different cases: a sleeping period of 1 second and 2 seconds. Discuss the results.

FURTHER READING

Rappaport's book [8] is an advanced reading about the physical properties of channels and wireless communications at their lowest level. A good idea is also to explore how exactly a modern physical layer works and what tricks are used, e.g., in the physical layer of IEEE 802.15.4. The full description of the standard can be found at the IEEE Standards webpage [5]. The non-uniform nature of wireless links was analyzed in the work of Zhou *et al.* [10]. An interesting piece of research is the work of Whitehouse *et al.* [9] on the so-called capture effect, which is able to differentiate between packet loss and collisions, and thus gives more information to the MAC protocols to mitigate the problem of collisions.

A very good survey of modern MAC protocols for WSN is the work of Huang *et al.* [4]. The original descriptions of the presented MAC protocols can be found in the remaining publications.

[1] L. Bernardo, R. Oliveira, M. Pereira, M. Macedo, and P. Pinto. A wireless sensor mac protocol for bursty data traffic. In *IEEE 18th International Symposium on Personal, Indoor and Mobile Radio Communications (PIMRC)*, pages 1–5, Sept 2007.

[2] Michael Buettner, Gary V. Yee, Eric Anderson, and Richard Han. X-MAC: a short preamble MAC protocol for duty-cycled wireless sensor networks. In *Proceedings of the the 4th ACM Conference on Embedded Networked Sensor Systems (SenSys)*, pages 307–320, 2006.

[3] Prabal Dutta, Stephen Dawson-Haggerty, Yin Chen, Chieh-Jan Mike Liang, and Andreas Terzis. Design and Evaluation of a Versatile and Efficient Receiver-initiated Link Layer for Low-power Wireless. In *Proceedings of the 8th ACM Conference on Embedded Networked Sensor Systems*, SenSys '10, pages 1–14, New York, NY, USA, 2010. ACM.

[4] Pei Huang, Li Xiao, S. Soltani, M. W. Mutka, and Ning Xi. The evolution of mac protocols in wireless sensor networks: A survey. *Communications Surveys Tutorials, IEEE*, **15**(1):101–120, First 2013. ISSN 1553-877X. doi: 10.1109/SURV.2012.040412.00105.

[5] IEEE. IEEE 802.15.4 Standard Specification, 2014. URL http://standards.ieee .org/about/get/802/802.15.html.

[6] David Moss and Philip Levis. BoX-MACs: Exploiting Physical and Link Layer Boundaries in LowPower Networking. Technical report, Stanford University, 2008.

[7] Venkatesh Rajendran, Katia Obraczka, and J. J. Garcia-Luna-Aceves. Energy-efficient Collision-free Medium Access Control for Wireless Sensor Networks. In *Proceedings of the 1st International Conference on Embedded Networked Sensor Systems*, SenSys '03, pages 181–192, New York, NY, USA, 2003. ACM. ISBN 1-58113-707-9.

[8] T. S. Rappaport, S. Y. Seidel, and K. Takamizawa. Statistical channel impulse response models for factory and openplan building radio communicate system deisgn. *IEEE Transactions on communications*, **39**(5):794–807, 1991. ISSN 0090-6778.

[9] K. Whitehouse, A. Woo, F. Jiang, J. Polastre, and D. Culler. Exploiting the capture effect for collision detection and recovery. In *Proceedings of the 2Nd IEEE Workshop on Embedded Networked Sensors*, EmNets '05, pages 45–52, Washington, DC, USA, 2005. IEEE Computer Society. ISBN 0-7803-9246-9.

[10] G. Zhou, T. He, S. Krishnamurthy, and J. A. Stankovic. Models and solutions for radio irregularity in wireless sensor networks. *ACM Transactions on Sensor Networks*, **2**(2):221–262, 2006.

4

LINK MANAGEMENT

This chapter focuses on how wireless links are managed and evaluated. Different from wired links, where the beginning and end of the wire unambiguously connect two or more nodes and thus guarantee a working communication link between them, wireless links use the shared wireless medium. Links between nodes come and go, and interfere with each other without any possibility for control. In this chapter, you will learn what techniques have been developed to discover and manage links as well as how to evaluate their quality. If you are already proficient in the area of link management and quality estimation, you can quickly skim this chapter and jump to the Summary and Questions and Exercises sections.

4.1 WIRELESS LINKS INTRODUCTION

The previous chapter explored wireless communications and their properties. It also discussed the problems associated with wireless communications, such as noise and interference and the practical problems arising from them, e.g., the hidden terminal problem or exposed terminal problem. At the level of MAC protocols, all we need is to understand when a node is allowed to access the shared wireless channel, without disrupting the work of others. The MAC protocol, however, does not guarantee that the transmission will be successful and also does not try to guarantee it. In some cases, the MAC protocol uses acknowledgements to retransmit a packet. However, the MAC protocol cannot say how often it will need to retransmit the packet in order

Introduction To Wireless Sensor Networks, First Edition. Anna Förster.
© 2016 The Institute of Electrical and Electronics Engineers, Inc. Published 2016 by John Wiley & Sons, Inc.

to succeed. This is the task of link layer protocols, which abstract the network as a set of links between individual nodes and attempt to characterize those links in terms of their quality or reliability. Link quality protocols pass this information to other protocols such as routing protocols.

Moving away from technical details, the following summarizes what every user and developer of sensor networks needs to be most aware of. The main message is:

> A wireless link is unreliable, not symmetric, and highly fluctuates in time and space.

First, we need good metrics to characterize wireless links and their reliability and quality. Typical metrics are:

- **Packet Reception Ratio or Rate (PRR).** This is the ratio between successfully delivered packets and all sent packets. PRR is measured in percentage and easily computable on any node without needing any special hardware.
- **Received Signal Strength Indicator (RSSI).** This is the signal strength, expressed in dBm (decibel per meter), for each received packet. This information is provided by the radio transceiver itself, but is often encoded, e.g., a node provides the results as a number, which later needs to be converted to dBm. The higher the RSSI, the better the signal. However, typical signal strengths are negative.
- **Link Quality Indicator (LQI).** This is a score given for each individual packet by the radio transceiver. It is part of the IEEE802.15 standard, but its implementation is different for various vendors. In general, it is a positive number ranging from approximately 110 to 50, where higher values indicate better quality.
- **Signal to Noise Ratio (SNR).** This is the difference between the pure signal and the background noise, given in dB.

Measuring the last metric, SNR, is not trivial in many platforms. The user needs to first measure the RSSI of the noise itself, next measure the RSSI of a packet, and then subtract them. The first metric, PRR, is evidently the most concrete and practical one. It tells you immediately and in a readable way how good a particular link is. The problem with this metric is that you need to send many packets over the same link to calculate it reliably. Furthermore, it can change over time and space, which requires regular recalculation. Sometimes the metric is simply there, e.g., for links currently in use. However, it is missing for the links you are not currently using and it can be very costly to obtain. Thus, researchers have tried to identify some relations between PRR and other, simpler metrics such as RSSI, LQI, and SNR.

Section 4.2 discusses in greater depth wireless links' properties, using some of the preceding metrics. Afterward, all tasks of the link layer protocols are individually explored.

4.2   PROPERTIES OF WIRELESS LINKS

Wireless links are highly unpredictable and time-varying phenomena. The study of their properties could fill whole libraries and the details are highly scientific. The following is a summary of the most fundamental and relevant properties, which will help us manage and understand our sensor networks more efficiently.

4.2.1   Links and Geographic Distance

The first question to explore is how path loss correlates with distance between the communicating nodes (recall the definition of path loss from Section 3.2). Figure 4.1 demonstrates the complexity of this correlation. Here, a simple experiment has been performed with sending data packets from a sender to a receiver at different distances between the two of them. The packet reception ratio is depicted against the distance between the nodes. If the world were simple, the signal should have gradually attenuated with growing distance from the sender. It can be seen that with a very long distance between the communicating nodes, the PRR drops so low as to not allow for communication at all. However, at short distances, the outcome of the communication is completely unpredictable: sometimes it works very well, sometimes it does not work at all. All processes explained in Section 3.2 affect the signal and the result is highly unreliable communication.

To bring some order into the chaos, Woo *et al.* [8] started differentiating between three types of communication zones.

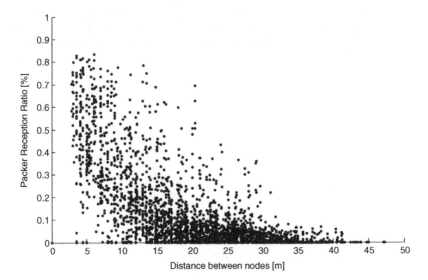

FIGURE 4.1   Packet reception ratio between a single sender and a receiver with different distances between them. The used dataset was gathered in 2004 at the Massachusetts Institute of Technology at their Intel Lab [2].

- **Effective communication zone.** In this zone, nodes communicate with at least a 90% delivery rate.
- **Transitional communication zone.** In this zone, everything is possible, from almost 100% to 0% delivery rate. Some links might be stable, whereas others might emerge and disappear quickly.
- **Clear communication zone.** In this zone, communication is almost impossible, but interference might still occur.

Looking at Figure 4.1, it becomes clear that this differentiation is ideal. None of the links lie in the effective communication zone. This comes from the environment itself and from high interference. The zones are also not necessarily geographically well defined, but they do quantify the link quality between certain nodes in the networks. This gave rise to the idea of measuring link quality, in which the quality is simply identified as good, intermediate, or bad. Some of the link quality protocols presented next use this approach to present their results to higher layers such as routing.

4.2.2 Asymmetric Links

Figure 4.1 shows that distance does not dictate communication success directly.

Thus, we also need to ask the question of whether there is any guarantee that if communication works in one direction it will also work in the opposite direction? symmetric Such links are called *symmetric links*. This means, if node A can send packets to links node B, then node B is able to send packets to node A. Thus, the link between A and B works in both directions or can be defined as symmetric.

However, there is no guarantee for this since transmission ranges are rarely circular. Sometimes node B cannot send any packets to node A or only a few are success-asymmetric fully transmitted. This causes *asymmetric links*. Figure 4.2 illustrates how asymmetric links links occur.

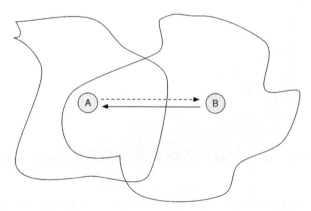

FIGURE 4.2 Asymmetric links caused by irregular transmission ranges. Node B can reach node A, but A cannot reach B.

Asymmetric links are extremely problematic for wireless communications because they are hard to spot. For example, when node A receives something from node B, it tries to answer. Node B does not receive an answer, so it sends the request again. Node A assumes this is a new request and sends something again, and so on. The case of perfectly asymmetric links is easier to handle if no packets get from node A to node B because node B will give up at some point. However, if some of the packets are received and some not, then both nodes will keep up the connection instead of giving it up and searching for a better one. Unfortunately, the latter case of half-asymmetric links is the rule.

4.2.3   Link Stability and Burstiness

Along with the preceding problems, there is also *burstiness*. Scientifically speaking,   **burstiness**
burstiness is a property of wireless links which manifests itself in the fact that wireless links do not have an independent probability of failure. If a failure happens, it happens for a continuous amount of time and not punctually.

Very often the burstiness of wireless links is measured with a *conditional prob-*   **CPDF**
ability delivery function or CPDF function. Examples of the function for an ideally   **function**
bursty link, a linear bursty link, and an independent link are depicted in Figure 4.3. The graph and the CPDF function reads as follows: If a node sends two packets to another node, the probability of receiving the second packet depends on whether the first packet was received. If the first packet was received on an ideally bursty link ($+1$ on the x axis), then the probability of receiving the second packet is b. If it was not received (-1 on the x axis), then the probability of receiving the second is c, where $c < b$. The function expands into both directions on the x axis to cover cases such as "when X packets were successfully received, then the next packet ...". With independent links, the probability for the next packet reception does not depend on whether the previous packet was received or not. In this case, $b = c$.

The burstiness is rarely ideal and in reality resembles something more like the linear bursty link in Figure 4.3 or a combination of an ideal and a linear bursty link. In this case, the probability of receiving the next packet grows with the number of successfully received packets and vice versa.

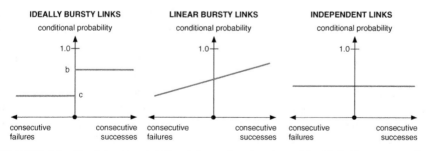

FIGURE 4.3   Conditional probability delivery function for an ideal link, linear bursty link and an independent link.

Burstiness has been extensively explored in wireless sensor networks research such as in Srinivasan *et al.* [6] and Lertpratchya *et al.* [5]. Burstiness is often observed in the transitional communication zone, as defined in Section 4.2.1.

4.3   ERROR CONTROL

Errors happen all the time. However, you need to control these errors, recognize them, and even prevent them to enable efficient applications. Error control is a very important concept but often neglected by researchers. Besides some basic approaches, it is rare to see a real-world application with fully implemented error control. The main purpose of error control is to guarantee communication is **error-free, in-sequence, duplicate-free, and loss-free.**

These properties are adopted from Internet-based communications, in which packets are parts of frames and their sequence is important. Sensor networks work in a slightly different way and some of the previous requirements are largely relaxed. For example, in sequence does not play a very important role (unless the sensor network is transporting life media streams, which is rather rare) because packets are self-sustainable and can easily be put back in sequence given their creation timestamps. Duplicates are not welcome, but also not damaging, as they can be also easily recognized and deleted at the sink. What remains are these two requirements:

- **Error-free.** Errors result either in faulty sensor data, which is harmful to the application, or in resending of packets, which is harmful to the network lifetime and throughput.
- **Loss-free.** Losses are harmful to the application and might render the whole network useless.

In terms of sensor networks, the following discussion focuses on these two types of problems. In general, backward error control and forward error control are the two types of error control schemes that are used.

4.3.1   Backward Error Control

Backward error control refers to the fact that you should not try to prevent errors but only discover them. Once we discover a problem with a particular packet received, we should send the so called Automatic Repeat Request (ARQ) to the sender. How to discover the error? The simplest way is to use a Cyclic Redundancy Check (CRC). This is a function computed over the contents of the packet and attached to it. For example, you could compute the sum of all bits in the packets, often referred to as checksum   the *checksum*. Figure 4.4 provides an example of a packet with a computed checksum and an error discovery on the receiver's side. In this case, we have used the simplest parity bit   CRC checksum, the *parity bit*. The parity bit computes whether the number of 1s in the packet is even or odd, marking an even number with 0 and an odd number with 1. This is the same as an AND computation over the bits.

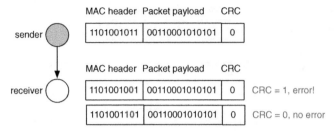

FIGURE 4.4   Computation of CRC and error detection of a sample.

Of course, the parity bit is simple and does not detect all errors. For example, in the first shown packet for the receiver in Figure 4.4, there is only one error, transforming a 0 into 1. This error is correctly detected by the parity bit. However, the second packet has two errors, which neutralize each other. The computed parity bit corresponds to the received one, thus no error is detected. For this reason, more complex CRC code computations with more bits exist. The interested reader can find more information about them and how to compute them in [7].

Rarely will you need to implement CRC codes, since they are already included in many protocols such as 802.15.4 (the physical and MAC layers of Zigbee were discussed in Chapter 3). In fact, CRC computation and checking is even implemented in hardware for most of the typical radio transceivers for sensor networks, such as the CC2420, also part of our Z1 sensor node.

4.3.2 Forward Error Control

Forward error control (FEC) attempts to prevent errors instead of only detecting them. This is reasonable, because

> Sending a new packet is very costly and uses not only energy, but also precious bandwidth. Sending a couple of more bits or bytes into a packet, which needs to get out anyway, is almost without cost.

This is due to the overhead involved in sending a single packet. Recall MAC protocols from Chapter 3. They involve a lot of administrative tasks before you can actually send the packet out, like the long preambles of BMAC or the CTS/RTS handshaking of CSMA/CA. Thus, it is a good idea to avoid resending a packet if you can prevent errors by adding more information into the first packet.

The simplest FEC technique is to repeat the packet payload several times in the same packet, called *repeating code* and shown in Figure 4.5. Here, the payload can be *repeating code* reconstructed if not too many errors happened to one of the repeated bytes. However, what happens if the link is bursty, as often observed? Recall from Section 4.2.3, that when errors happen on a link they are typically not random over the individual bits but clustered for some continuous number of bits. Thus, you can expect that the errors are similar to the second example in Figure 4.5, in which two copies of the *c* character are lost and cannot be reconstructed.

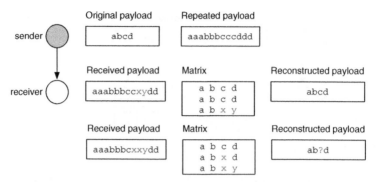

FIGURE 4.5   A FEC scheme with repeated individual payload bytes.

interleaving   The solution to this problem is called *interleaving*. The idea is simple, but powerful. Instead of repeating individual bytes, repeat the whole payload, as shown in Figure 4.6. Thus, when a bursty error happens, it involves different bytes but never the same one. Of course, there are many more FEC techniques in the literature [4].

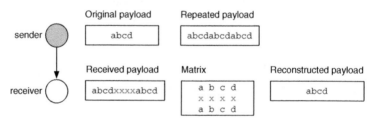

FIGURE 4.6   A FEC interleaving scheme with repeated individual payload bytes. With interleaving, bursty errors are better handled.

4.4   NAMING AND ADDRESSING

One question to resolve before moving on to link quality protocols is how to address individual nodes. Fundamental differences are made between naming and addressing. While naming resembles more human names, which are usually kept forever, addressing is more like your current home address.

4.4.1   Naming

Naming gives some abstract identification to individual nodes. No real restrictions apply, but typically it is practical to use unique names in the same network and to represent them in an efficient way. The most common way to name nodes in a sensor network is to give them IDs, which are represented by integers and are unique.

It is very important to understand that a name does not reveal anything about the position of the node. However, it can reveal something about the role of the node (e.g., sink or gateway) or the type of data it can deliver. An extreme case of the latter is the so called *data naming*, where nodes with the same type of sensors share also the same name. For example, all nodes with temperature sensors have ID 1 and all nodes with light sensors have ID 2. This is sometimes useful, particularly when the application does not care where the data comes from, but only what kind of data is delivered. **data naming**

4.4.2 Addressing

Different from naming, addressing reveals more about the position of the node than its role or abilities. A typical example of addressing is IP addresses, which clearly identify the position of individual nodes given their addresses. In sensor networks, IP addresses are rarely used besides maybe for the sink or gateway of the network. More often, the address is the direct geographical or relative position of the node.

4.4.3 Assignment of Addresses and Names

For small networks, it might seem trivial to assign unique names or addresses to nodes. You could just take IDs 1, 2, 3, for names and the real GPS positions of the nodes as their addresses. However, both become important with larger networks or with randomly deployed networks. Protocols and algorithms for automatic name and address assignment are out of scope for this book, but the following should be kept in mind:

- **Size and length.** You should plan for scalability and not rely on the fact that the network will always remain small. The transition from IPv4 to IPv6 addresses is an example of a problem this thinking could create later on. IPv4 uses 32 bit addresses, which corresponds to a maximum of 4.3 billion addresses. With the explosion of the Internet of Things, this number became too small and IPv6 addresses were introduced with 128 bits, capable of representing an amazing 3.4×10^{38} addresses or 40,000 addresses for each atom on Earth. With sensor networks, you should not go that far but reserving several bytes might be better than a single one.
- **Storage.** The next problem is where to store the address and the name. If stored in the normal volatile memory, the node will loose the information every time it reboots.
- **Positioning and repositioning.** Another critical point is how to identify the address of a node and how to update it if the node moves away. Chapter 8 addresses this problem.
- **Uniqueness.** Even if some applications permit reuse of names, it is typically important to have unique names in the same network. For example, this is easy to achieve when new nodes are simply given incremental IDs. This is also the

reason why a large count of unique IDs is important. Another very common option is to give the nodes their own fabrication numbers or MAC addresses as IDs. These are, however, typically very long.

4.4.4   Using Names and Addresses

Names are typically too abstract whereas addresses are too exact. Thus, in sensor networks a combination is very often used, where the name (ID) of the node points to the identity of the node and some of its properties or abilities and the address represents its current position (relative or absolute). For some purposes, such as link quality prediction (which are explored next) only the name is really relevant. For other protocols, such as routing, the addresses are more relevant and the ID becomes less important.

4.5   LINK ESTIMATION PROTOCOLS

In the previous sections, we have seen how unreliable and complex wireless links are. However, we still want to use them but you need to use them efficiently. Thus, we need to find a way to identify good links. This is the job of a link quality protocol, sometimes also called *link estimation* protocol or *neighborhood management* protocol.

link
estimation
neighborhood
management

4.5.1   Design Criteria

The main task of a link management protocol or simply a link protocol is to identify good communication partners with stable links. It typically serves another protocol, such as routing or clustering, and provides it with this information as needed. A good link protocol complies with the following criteria, which often stand against each other:

- **Precision.** The protocol provides precise and correct results, i.e., if it says the link is good/bad, then the link really is.
- **Agility.** The protocol reacts quickly to changes in the links properties, i.e., if the link becomes bad, the protocol quickly recognizes and signals it.
- **Stability.** The protocol does not fluctuate between decisions, i.e., if a link becomes bad for some short period of time and then recovers again, the protocol sticks to the more time-invariant quality. Obviously, stability is the opposite of agility and a tradeoff needs to be found.
- **Efficiency.** The protocol achieves the preceding properties at very low resource cost, such as energy, memory, or processing. Again, this property is in conflict with all the previous so a tradeoff is needed.

We now have all necessary tools and knowledge to actually identify good link estimation protocols. In the following sections, we will see some fundamental options of how to estimate the quality of individual links with various metrics and we will see one concrete protocol called the Collection Tree Protocol.

4.5.2 Link Quality Based

Direct and indirect evaluation are two options we have to evaluate a wireless link. Direct evaluation refers to some measurement of the link quality such as RSSI, LQI, etc. This value gives a lot of information about the link between two nodes. However, earlier in this chapter, you learned these metrics do not necessarily correspond to whether communication is possible or good. Thus, the indirect method simply measures how reliable communication between the nodes is, without paying attention to the physical properties of the link. The following discussion starts with the direct method.

To measure the link quality directly, you need to send several packets (not too many) and to compute the mean of the measurements. For example, assume you have one node in the middle of the network, which sends the packets and the other nodes receive them and answer. If you use RSSI and LQI as metrics each neighbor has to store the RSSI and LQI of the packet it received from the central node and to send this information back. When you receive this information, you can extract the RSSI and LQI of this packet and the packet itself tells you what the observed quality was at the other side.

Now, you need to translate the RSSI/LQI values into a link estimation. Typically you should use thresholds to do this, e.g., all links with RSSI > −60 and LQI > 95 are considered very good. Is this method a good one to estimate the link quality? Let us explore the criteria we defined in Section 4.5.1:

- **Precision.** The precision of this method is not very good. The reason being, RSSI and LQI do not translate well into communication success, which is our final goal.

- **Agility.** When something changes, for example, a new piece of furniture is placed between us and one of your neighbors, this method will rather quickly identify the change, because the RSSI value will most probably change immediately. However, there is still no guarantee that the communication success rate actually changed.

- **Stability.** Quite the opposite, if something changes only very shortly, e.g. a human passes between two nodes, this method will not be able to filter this fluctuation away.

- **Efficiency.** Here you need to split the discussion in terms of the different resources involved. In terms of communication resources, the method is quite efficient, as it needs one or few packets to identify the quality. The same goes for memory, because only one or few values are stored. However, it needs some processing resources to translate the measured RSSI/LQI into communication quality, to compare with thresholds, etc.

4.5.3   Delivery Rate Based

To measure communication success and delivery rates, we need to communicate. For example, you could send 10 or 100 packets to all your neighbors and wait for their response to see who received the packets and how many. For example, assume one node in the middle of the network, sending out packets and searching for communication partners. Assume also that you have sent 100 packets to whoever can receive them and you start receiving acknowledgments from your neighbors. The first question is whether you can reliably receive all of them, since all neighbors are sending at the same time. Probably not but let us explore this question a little further.

If you send the packets to individual nodes, you automatically make the assumption that you know who is out there. However, this is not the task of your protocol. You need to discover who is out there and how well you can communicate with them. You need to find another possibility to solve the problem. For example, a good MAC protocol can make sure that in the same communication neighborhood, no two transmissions are taking place simultaneously or, at least, not too many. Question 4.4 at the end of this chapter discusses alternatives and is left to the interested reader as a homework.

Now that we have solved this issue, we are ready to continue with link quality estimation. You can count the number of received acknowledgements to estimate communication success. For example, you have received only 12 ACKs from node 3. This gives you a communication success rate of 12/100 or only 12%, which is not great. Other nodes, such as 2 and 4, have much better success rates, 98% and 99%, respectively.

Next, you need to keep track of all communication to all neighbors and update the previous success percentages.

Where do the differences come from? Perhaps different distances? Or extreme path loss somewhere in the environment? You do not need to care. All you need to care about is that the link works. Why then did this chapter look at the direct link quality measurement, if the indirect method seems to be much more direct and much more successful? Because the indirect one also has some disadvantages. Let us further explore the design properties that were defined in Section 4.5.1.

Precision.   The precision of this method is quite good, as it always gives the currently known success rate for a neighbor.

Agility.   When something changes, perhaps a neighbor disappears completely, you may need quite a lot of time to actually spot this change. Recall that all you are doing is recalculating the success rate when a new packet is sent/acked. When you loose one packet, you can correct your success rate slightly, but not significantly. You may have to loose many packets before the success rate starts to look really bad and the neighbor is not used any more. Of course, the agility of this method depends on the *recalculation window* you use. For example, do you calculate the success rate over the last 100 packets or the last 5 packets? The smaller the window (5 packets), the more agile the method will be.

recalculation
window

Stability. Here you have the opposite situation than that for agility. If you use a large recalculation window, your method will be quite stable, as small fluctuations will not be taken into account. Vice versa, if the window is small, your method will be highly agile but unstable, as even one lost packet will let your success rate drop dramatically. You need a medium-sized window to be a little agile and stable at the same time. Windows with 10 to 20 packets work just fine.

Efficiency. Question 4.5 in the Questions and Exercises section addresses this topic as homework.

In summary, the direct method better predicts what can be expected from a particular link in the future. However, this quality is not necessarily the same as the communication success and can be misleading. The indirect method, which measures delivery rates instead of wireless properties, gives more precise results about the communication success rate and recalculation window, but it is not able to predict what may happen in the future. A combination of both can also be used.

4.5.4   Passive and Active Estimators

Besides the used metric and required properties of link quality protocols, there are two basic ways to implement them: passively and actively.

Passive estimators do not produce any packets by themselves. Instead, they start with all known neighbors with the same mean quality and let other protocols and algorithms on the node create, send, and receive packets. From this *passive snooping*, they learn about the properties of the link. This implementation type is very resource efficient and stable, but performs rather poorly in terms of agility and precision. The main problem is that the goals of the link estimator are different from the goals of other protocols on board, which are not interested in ever using bad neighbors. Thus, if new neighbors arrive or bad neighbors become good ones, the passive snooper has little chance of spotting the change.

passive snooping

Active estimators take over the control and produce packets when considered needed. For example, if the estimator has not heard back from a particular neighbor for some time, it creates a request and sends it to that neighbor to update its information. This type of protocol is precise, agile, and can also be stable. However, they are not resource efficient, since they also maintain information, which might never become important. For example, if the only service user of the link estimator is a routing protocol, which always sends everything to a single sink, then all neighbors further away from the sink will remain irrelevant throughout the network lifetime. However, the link estimator is not able to identify this and will continue maintaining up-to-date information.

4.5.5   Collection Tree Protocol

This section discusses a concrete link quality protocol, the Collection Tree Protocol (CTP) by Gnawali *et al.* [3]. CTP is a combination of a routing protocol with integrated link estimator. In terms of routing, it belongs to the family of converge cast protocols, which route everything to a single sink by means of a tree structure. You can imagine the network as a graph, where a tree is built rooted at the sink and spanning all nodes in the network, such that the routing costs from all nodes to the sink is minimal. Chapter 5 looks at CTP again from the routing perspective whereas the following discussion focuses on its link estimation part.

ETX   To understand how link estimation works, you first need an overview of CTP's algorithm. Every node maintains a list of all known neighbors with their *ETX* estimations. ETX is similar to PRR and stays for Estimated ReTransmissions and gives the estimated number of retransmissions to reach that neighbor. For example, if a particular neighbor has an ETX of 1.5, this means that you need to retransmit 1.5 times (statistically) to reach it. ETX can be used in terms of reaching a direct neighbor or can be summed up to give the expected retransmissions to reach a remote node. In CTP, the ETX measure is linked to reaching the sink of the network through the best currently available route.

Data packets are sent to a neighbor directly and the neighbor is selected according to its ETX count (the lowest option is the best). Additionally, CTP sends control beacons (link estimation active packets) to check the status of other neighbors. This makes CTP an active link estimator. Furthermore, it is an adaptive one, because it sends the control beacons not always at the same rate, but changes the sending frequency according to the current network situation. Control beacons are simply tiny packets without any real data inside, which contain the sender's currently best ETX to the sink. This helps the sender's neighbors to calculate their own best ETX.

CTP defines a special sending interval of τ_{low}. This is the lowest frequency sending period for the control packets, when nothing really interesting happens in the network. However, if one of the following situations occur, then the period is reset immediately to τ_{high}, which sends many more control beacons:

- A data packet is received from a neighbor, whose ETX count is not greater than your own. This means that your neighbor, or maybe many of them, have old information and you need to send them an update.
- Your own ETX decreases significantly (e.g., with more than 1.5 ETX in the original version of CTP). Thus, you have much better costs than before so you need to propagate the good news to your neighbors.
- A beacon with its P bit set to 1. The P(Pull) bit is a special bit in the beacon, which signals that the sender is new or has lost all its ETX estimates and needs to update its information as soon as possible. This will trigger higher control beacons frequency at all neighbors.

When one of the previous situations occur, the link estimator switches to its maximum beacon frequency, given by τ_{high}. Then it exponentially increases the period between the beacons again, up to τ_{low}.

How well does the protocol perform? Well enough for most applications. It is fairly precise, agile, and the threshold of 1.5 ETX also gives it some stability. Some careful parameterization is needed to avoid wasting too many resources. Some optimizations are possible, left for homework discussion.

4.6 TOPOLOGY CONTROL

Until now, discussion has focused on how to measure existing wireless links as well as their properties and quality. However, you can also control these properties, for example, by increasing and decreasing the transmission power.

The higher the transmission power, the more energy you use and the better the link becomes. There is always some maximum (e.g., 0 dBm for Z1 nodes) and typically several transmission levels, which you can select (8 levels for Z1, from −24 dBm to 0 dBm).

The lower the transmission power, the more restricted the transmission area is. This is not necessarily a disadvantage. Very dense networks often suffer from overloading and too many neighbors.

Allowing each node to use its own transmission power poses several communication challenges:

- **Asymmetric links** become the rule, as two communication nodes are allowed to use different transmission powers.
- **Directional links** become possible, but are rather undesired. While in theory nodes can save energy by using only the minimally required transmission power to send data traffic and avoid any acknowledgements or control beacons, this often renders the whole system fragile and unreliable. If something goes wrong, the sending node has no chance of discovering the problem.
- **Discovery of neighbors** becomes challenging because they could have too low transmission power to answer a query.

Given the preceeding issues, it is rather rare to see a topology control protocol that allows for a completely free choice of transmission power and especially selfish decisions. Typically either unified transmission power is selected for all nodes or a transmission power is selected so that communication neighborhoods can successfully talk to each other. Apart from these options, there are two possibilities of how to agree on the transmission power: centralized or distributed.

4.6.1 Centralized Topology Control

In centralized protocols, one of the nodes takes a leading role in the process and disseminates its decision to all other nodes, which have to obey. This is typically a more important or powerful node, such as the base station or the sink. In order for this approach to work, the nodes have to send some communication information to the sink to analyze and take decisions. There are two major options:

- The nodes try out with their current transmission power and store the received RSSI and LQI values. They send them to the sink and the sink analyzes them to decide whether to lower the transmission power and by how much.
- The sink first sends a command to all nodes to change their transmission power and to report on the new available neighbors.

Typically, a combination of both approaches is used, in one or another order. However, the major problem comes from the fact that if the sink commands the node to go to a low transmission power, they might not be able to report anything. They are only able to receive a new command later on (the transmission power change affects only the sending process, not the receiving one). Of course, the sink could decide also for different transmission powers for individual nodes, but this is typically avoided because of too high complexity.

Centralized control is usually simpler, but it also requires significant communication overhead and does not react well in case of changes.

4.6.2 Distributed Topology Control

In distributed control, the optimal transmission power is adjusted at individual nodes without having somebody to control or oversee the process. Nodes try out different transmission powers and evaluate their neighborhood. The greatest challenge here is to synchronize more or less the process at neighboring nodes so that neighbors can understand each other.

For example, node 1 and node 2 try to find their minimal transmission powers so that they can still communicate to each other. Remember that they could be different because of asymmetric links. Node 1 incrementally decreases its transmission power and every time it does so, it sends a request to node 2 to check whether it can still reach it. If node 2 does not play currently with its transmission power, this approach would work well and node 1 will soon discover the transmission power at which it can easily reach node 2. However, if node 2 is also playing around while replying to node 1's requests, the situation becomes tricky. If node 1 does not receive the reply, what is the reason? That its own transmission power is too low and its request was not received at all at node 2? Or because node 2's transmission power is too low?

The solution to this problem is to allow only for maximum transmission power when sending replies. Thus, the nodes can play simultaneously with their transmission powers but when they answer requests from other nodes, they have to switch to the maximum.

Distributed control is more flexible, situation-aware, and precise than centralized control. However, it is also complex and it takes some time before all nodes stabilize at their preferred transmission power.

LINK MANAGEMENT: SUMMARY

In this chapter, you learned that wireless links are highly unreliable and their properties severely fluctuate with time and space. You can measure wireless link quality with several metrics, such as *packet reception rate* (PRR), *received signal strength indicator* (RSSI), *link quality indicator* (LQI), *or signal to noise ratio* (SNR).

⇒ Quality is not linear with geographic distance.
⇒ Asymmetry exists, where quality is different in both directions.
⇒ Wireless links are bursty, i.e., bit errors occur not independently.

Link management involves several tasks:

- **Error control** differentiates between *forward error control* and *backward error control*. The first tries to avoid errors and the second tries to repair errors that already occurred.
- You assign unique **names** to individual nodes, which refer to their role or properties, and **addresses**, which refer to their positions.
- **Link estimation** refers to maintaining quality indicators for individual neighbors. It can be delivery rate based or link quality based. The first option uses statistical delivery success between neighbors, the latter uses metrics such as LQI or RSSI to predict the communication success. Link estimation can be *passive* (overhear existing communication only) and *active* (creating its own communication). CTP uses a delivery rate based link estimator. Link estimation has to fulfill the following properties (often in conflict with each other): *precision, agility, stability,* and *efficiency.*
- **Topology control** plays with the transmission power of individual nodes to change their communication neighbors. It can be designed to be *centralized* (one node decides about the transmission power of all) or *distributed* (each node decides on its own). The first is simpler to implement and faster to converge, while the latter option is more flexible, but takes time to converge.

All tasks of the link management layer can be implemented as independent protocols or can be interleaved with each other.

QUESTIONS AND EXERCISES

4.1. Repeat the experiment shown in Figure 4.1 at very short distances between the nodes, e.g., 0.05 to 1 meter. What do you observe? Discuss the results.

4.2. Recall the FEC simple repeater scheme from Section 4.3.2 and Figure 4.5. What is the maximum number of errors per byte (char), so that the payload can be successfully reconstructed at the receiver? Assume each byte is repeated N times and express your answer in terms of N.

4.3. Recall the FEC interleaving repeater scheme from Section 4.3.2 and Figure 4.6. What is the maximum length of a bursty error in bytes (chars), so that the payload can be successfully reconstructed at the receiver? Assume the payload is repeated N times and express your answer in terms of N.

4.4. Recall the problem of simultaneously answering neighbors from Section 4.5.3. As discussed, a MAC protocol could help you out of the dilemma and make sure that no two transmissions are taking place at the same time. However, this will not always work because of the hidden terminal problem (see Section 3.2.2). Discuss other alternatives that you can discover in terms of new neighbors and at the same time avoid their simultaneous answers and consecutive loss.

4.5. Recall the indirect link quality measurement method from Section 4.5.3. Is this method resource efficient? Is it more resource efficient than the direct method or not? Why?

4.6. Recall the link estimator of CTP from Section 4.5.5. The data packets include the current ETX of the sender, otherwise, you cannot know that the sender has illegal information. Discuss the possibility to switch off completely the control beacons when nothing happens. What other changes are necessary to make this work? Are you saving resources? Are you getting less or more agile, precise, or stable?

FURTHER READING

A very good technical introduction to the properties of wireless links is given in the survey of Baccour *et al.* [1]. A more practical survey of wireless links in the context of sensor networks is presented in Woo *et al.* [8]. This work also defines the idea of communication zones. Burstiness is explored in great detail in the works of Lertpratchya *et al.* [5] and Srinivasan [6].

Error control and CRC computation are explored by Howard *et al.* [4] and Williams [7]. All details about CTP are given in the original publication of the protocol by Gnawali *et al.* [3].

At the web site [2] you can find the Intel sensor dataset, which this chapter has used to explore the correlation between PRR and distance (Section 4.2.1).

[1] N. Baccour, A. Koubâa, L. Mottola, M. A. Zúniga, H. Youssef, C. A. Boano, and M. Alves. Radio Link Quality Estimation in Wireless Sensor Networks: A Survey. *ACM Transactions on Sensor Networks*, 8(4), September 2012. ISSN 1550-4859.

[2] P. Bodik, W. Hong, C. Guestrin, S. Madden, M. Paskin, and R. Thibaux. Intel lab sensor data. URL http://db.csail.mit.edu/labdata/labdata.html.

[3] O. Gnawali, R. Fonseca, K. Jamieson, D. Moss, and P. Levis. Collection tree protocol. In *Proceedings of the 7th ACM Conference on Embedded Networked Sensor Systems (SenSys)*, pages 1–14, Berkeley, California, 2009. ACM.

[4] S. L. Howard, C. Schlegel, and K. Iniewski. Error Control Coding in Low-power Wireless Sensor Networks: When is ECC Energy-efficient? *EURASIP Journal on Wireless Communication Networks*, 2006(2):29–29, April 2006. ISSN 1687-1472.

[5] D. Lertpratchya, G. F. Riley, and D. M. Blough. Simulating frame-level bursty links in wireless networks. In *Proceedings of the 7th International ICST Conference on Simulation Tools and Techniques*, SIMUTools '14, pages 108–117, ICST, Brussels, Belgium, Belgium, 2014. ICST (Institute for Computer Sciences, Social-Informatics and Telecommunications Engineering). ISBN 978-1-63190-007-5.

[6] Kannan Srinivasan, Maria A. Kazandjieva, Saatvik Agarwal, and Philip Levis. The beta-factor: Measuring Wireless Link Burstiness. In *Proceedings of the 6th ACM Conference on Embedded Network Sensor Systems*, SenSys '08, pages 29–42, New York, NY, USA, 2008. ACM. ISBN 9781-59593-990-6.

[7] R. N. Williams. The CRC Pitstop. URL http://www.ross.net/crc/.

[8] A. Woo, T. Tong, and D. Culler. Taming the underlying challenges of reliable multihop routing in sensor networks. In *Proceedings of the 1st international conference on Embedded networked sensor systems (SenSys)*, pages 14–27, Los Angeles, CA, USA, 2003. ISBN 1-58113-707-9.

5

MULTI-HOP COMMUNICATIONS

Wireless sensor networks consist of several dozens to hundreds of nodes. Chapter 4 discussed how neighboring nodes can communicate to each other and how to avoid problems and optimize the performance of these communications. This chapter concentrates on the next problem, which is how to exchange data between remote nodes, i.e. between non neighboring nodes. This is also called *routing* or *multi-hop communications* as well as *data collection*. This chapter explores several possible scenarios in terms of data collection and the most popular routing protocols currently in use.

routing
multi-hop
data
collection

5.1 ROUTING BASICS

There are several scenarios for multi-hop communication or routing. They are differentiated based on the source and the destination of the data inside the network. First, we will explicitly define source and destination, as well as some important special roles of the nodes in the network.

> **Definition 5.1.** *Data source. The node in the network, which produces the required data and is able to send it out to other nodes in the network.*

Introduction To Wireless Sensor Networks, First Edition. Anna Förster.
© 2016 The Institute of Electrical and Electronics Engineers, Inc. Published 2016 by John Wiley & Sons, Inc.

Definition 5.2. *Data destination.* *The node in the network, which requires the data and is able to receive it from other nodes in the network.*

Definition 5.3. *Data forwarder.* *Any node in the network, which is not the source or the destination of the data, which is able to receive it from another node and to send it further in the network.*

Definition 5.4. *Data sink.* *A dedicated node(s) in the network, which is the implicit destination of any data in this network.*

sink   The *sink* is typically a usual node in the network, which has either an additional broadband communication interface or is connected via a wire to a more powerful device (e.g., a laptop). While a normal data destination can change over time, the sink usually does not. At the same time, a sink can be mobile and change its position. Depending on which is the source and which is the destination of the data, you differentiate between four scenarios (Figure 5.1):

- **Full-network broadcast.** There is a single source of the data and all nodes in the network are implicit destinations of the data.
- **Unicast.** There is a single source and a single destination of the data, which could be any nodes in the network.
- **Multicast.** There is a single source, but multiple destinations of the data. Again, both the source and the destinations can be any nodes in the network.
- **Convergecast.** All nodes in the network are sources, while a single node (typically the sink) is the destination.

Full-network broadcast is typically used to inform all nodes in a network about something generally important for all of them such as new parameters or the position of the sink. Together with convergecast, it builds the backbone of multi-hop communications in sensor networks. Unicast and multicast are used to send data to some particular node(s) in the network. However, these scenarios are rather rare in real-world sensor networks.

Remark 5.1. *Convergecast communications could be mistaken with a set of unicast communications, i.e., between any source in the network and the sink. However, the data of convergecast communications are typically merged on the way, thus minimizing the traffic. Unicast communications are independent from one another.*

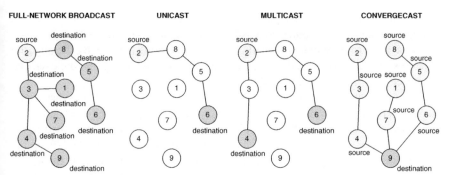

FIGURE 5.1   The various routing scenarios in a sensor network. Full-network broadcast delivers the message to all nodes in the network. Unicast has one source and one destination, which could be any nodes in the network. Multicast has several destinations. Convergecast is special for sensor networks and refers to the collection of data from all nodes to a dedicated sink.

Remark 5.2.   *The term* data collection protocol *is typically used in the context of convergecast scenarios. It refers to the fact that the sink is trying to collect all data from all sensor nodes.*

**data
collection
protocol**

Remark 5.3.   *Note that the problem of full-network broadcast sounds simple but in fact is complex. From the scenarios explored here, broadcast is the only one whose success* cannot *be guaranteed, as no node in the network really knows which nodes exist and which do not. Thus, it can be implemented only as a best effort, while the other scenarios can be forced to be guaranteed.*

**delivery
guarantee**

Let us now define routing:

> **Definition 5.5.**   *Routing is the process of selecting a sequence of nodes in the network, beginning from the data source(s) and ending at the data destination(s). This sequence of nodes is also called the routing path.*

A valid routing path exhibits some trivial, but important properties.

> **Definition 5.6.**   *A valid routing path consists of a finite number of nodes and it does not contain loops.*

The individual nodes on the path are also called *hops*, illustrating the fact that the data "hops" from one node to the next. There are many ways to find this sequence of nodes. It can be done *proactively* or *on-demand*, meaning the path was either found before data was available for routing or only after that. When done proactively, data

hops

**proactively/
on-demand**

can be sent over the ready to use path immediately, thus incurring very short delay. On the other hand, proactive routes are costly and might never be used, which increases the total cost of routing in the network. On-demand routing has exactly the opposite properties: the first packet is delivered with a typically large delay, but the overhead is incurred only in case of real need.

At the same time, routing paths can be determined centrally or in a distributed manner. Centrally determined paths are paths, which were computed or identified on some special node in the network (typically the sink) and then disseminated to the individual nodes for use. A distributed approach identifies the route directly on the individual nodes, taking into account their current environment, such as the neighboring nodes or available power. Thus, distributed approaches are able to cope better with failures or mobile nodes.

Remark 5.4. *In sensor networking applications, routing is always performed hop by hop. This means that no node in the network, not even the data source, is aware of the full routing path for a particular destination. Instead, only the next hop is known. This is different from IP or any address-based routing, where the complete path is determined before sending the data packet and strictly follows that path. A failure somewhere on the path would lead to a delivery failure and delivery restart. This is avoided in sensor networks with the concept of hop-by-hop routing.*

As you saw in the previous chapter, sensor networks do not use an addressing scheme to identify individual nodes. Thus, the main question is: How do you know where the destination is? The answer is that the destination typically informs the other nodes in the network where it is via a full-network broadcast (Section 5.3.1 discusses this exactly).

The next question is: How do you know which neighboring nodes are closer to the destination? The general answer is, you use routing metrics and some of the most important ones are discussed next.

5.2 ROUTING METRICS

The routing metric is the driving force behind routing. It represents a way to compare different sensor nodes in terms of their vicinity to the destination. This vicinity can be a precise or approximated geographic vicinity, but it can also be a communication-based vicinity such as least delay.

5.2.1 Location and Geographic Vicinity

Geographic location is probably the most logical metric to compare two nodes in terms of their vicinity to the destination. Real geographic locations can be used, but also approximations such as Manhattan[1] distances or similar. The main requirement is that all nodes are aware of their own position and that they have the position of

---

[1] Manhattan distance is the sum of the distances in each dimension. It is less costly to compute than normal Euclidian distance, but less precise.

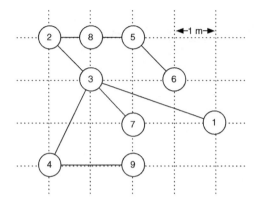

FIGURE 5.2   A sample network on a location based grid. While there is a correlation between distance and existing links between nodes, there is also no guarantee that close nodes will be able to communicate.

the destination (recall that this is communicated by the destination itself via a full-network broadcast).

The advantage of this metric is that it is clear and simple. Different next hops can be easily compared to each other. Furthermore, if the nodes are static, the metric is also static, meaning that it does not change over time. However, location awareness is also very costly. The nodes need either an additional location sensor (e.g., a GPS sensor) or a costly localization protocol to approximate the location.

Another disadvantage of location-based routing is that the metric does not take into account the communication network itself and its properties.

This is illustrated in Figure 5.2. For example, it looks like nodes 7 and 9 are really close to each other and can exchange messages directly. However, they cannot communicate and have to go through two other nodes for that (nodes 3 and 4). Thus, other metrics are usually preferred.

5.2.2   Hops

Another way of approximating location and vicinity to the destination is to use number of hops. Here, the assumption is that a hop corresponds to some distance in the network, thus more hops correspond to longer distance. This idea is visualized in Figure 5.3. Nodes that can directly reach the destination are one hop away from it; nodes that can directly reach any one-hop node are two hops away from the destination, etc. The number of hops is correlated with the real geographic distance but also takes into account the network topology and the available links in it. For example, node 5 in Figure 5.3 is closer geographically to the destination than node 9, but one hop further away than node 9. This is important, as geographic proximity does not guarantee the availability or the shortness of a route to the destination.

Hop-based routing exhibits some further advantages and disadvantages compared to geographic routing. First, the number of hops from the destination is not readily

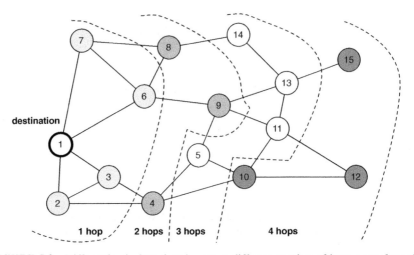

FIGURE 5.3   Differently shadowed nodes are at different number of hops away from the destination. Note that the number of hops are correlated with distance, but not always. For example, node 5 is even closer to the destination from a geographic point of view than node 9, but is one hop further away than node 9.

available at all nodes, but it needs to be discovered first. For this, usually the destination sends a full network broadcast to all nodes (refer to Section 5.3.1). During this broadcast, each node remembers from where it received the packet and how many hops away from the destination the node was. While rebroadcasting the packet, it will attach its own best available number of hops to inform its own neighbors, etc.

Hop-based routing is very simple to perform. The source sends the packet to the neighbor, which has the least hops to the destination, and then each next hop repeats this until the destination is reached. Hop-based routing is illustrated in Figure 5.4, where node 12 routes data to node 1. Note that there is almost never a single possible route. Instead, several neighbors of node 12 have a hop count of 3 and thus both are suited for routing. It becomes apparent in this example, that the metric used is not a routing protocol itself. A routing protocol is more than using a specific metric. It specifies the implementation details and further refines and optimizes the routing procedure.

Remark 5.5.   *Two protocols can use the same metric, but perform the routing differently and with different efficiency.*

5.2.3   Number of Retransmissions

In the best case scenario, the number of hops also represent how many times a packet needs to be forwarded to reach the destination. However, in reality, a packet often needs to be retransmitted several times before it reaches even the next hop. The main reason is interference, but also bad links over long distances or too many obstacles can cause packet loss and retransmission.

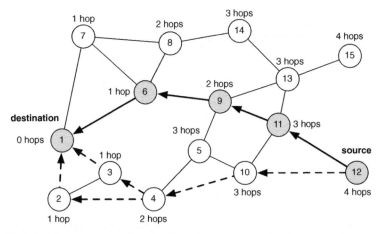

FIGURE 5.4   The same network as in Figure 5.3, where routing from node 12 to node 1 is performed. The selected route is bold black, while possible other routes are bold dashed. These routes are equivalent from the metric point of view. Which path will be selected depends on the implementation of the routing protocol.

You have already seen how to solve this problem in Chapter 4 in the definition of ETX or number of retransmissions. To better represent this behavior in the routing metric, you can use the real number of retransmissions over the route, instead of the hops.

Remark 5.6.   *The number of retransmissions ETX over a route is always at least the number of hops over the same route.*

While this metric is more precise in terms of the needed overhead to forward a packet, it is also harder to obtain and fluctuates a lot over time. This is due to the unreliable and bursty behavior of the links. Thus, usually statistical values over longer periods is used, not the last seen number of retransmissions.

5.2.4   Delivery Delay

The metrics already discussed are clearly good metrics, but not always helpful in a wireless sensor network. The delivery chance or speed depends not only on the topology of the network, but also on the underlying and overlying protocols used.

For example, recall TDMA medium access protocols from Section 3.3.2. A sample network with a sample TDMA schedule is provided in Figure 5.5. The nodes have statically assigned TDMA slots and each one is allowed to send a packet only in its slot. Thus, when node 12 sends a packet, it needs to wait until slot 6 to send it to the next hop, node 11. Node 11 receives it in the 6th slot, but needs to wait until its own slot 5 (hence a complete frame) to send it to node 9, etc. This schedule slows down the packet. However, the route through nodes 12, 11, 13, 14, 8, and 7 has a better

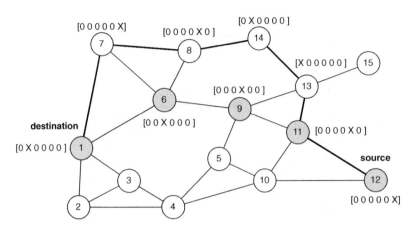

FIGURE 5.5   The hop-based shortest route is through the shaded nodes 12, 11, 9, 6, and 1. However, provided the TDMA schedule for the nodes in the network (not all are shown for simplicity), this route will take 16 slots for a packet to be delivered. The route in bold through nodes 12, 11, 13, 14, 8, 7, and 1 is one hop longer, but will take only 13 slots. Thus, shorter topological distances are not a guarantee for a shortest route in the network.

schedule. For example, when node 14 receives a packet from node 13 in slot 1, it can send it immediately out in slot 2. Thus, no time is wasted and the packet is "tunneled through" the network. This is similar to following a green wave of traffic lights on the street.

Delays can occur in a network for various reasons: TDMA schedules, application-layer processing or waiting, unreliable communications which need many retransmissions, etc. They can also come from propagation delays between the sending and receiving nodes, even if this delay is negligibly small compared to others sources of delay. Thus, it is essential to learn more about general delivery delay between any two nodes in a network:

Definition 5.7. *Delivery delay is the time elapsed between sending a packet at the source and receiving it at the destination, irrespective of the delay reason.*

Delay is a simple metric, which can be easily measured along the way from the source to the destination by time-stamping the packet at the source. However, it is not trivial to propagate this information back to the source for routing purposes, as it will need an additional packet to traverse the complete route.

5.3   ROUTING PROTOCOLS

A routing protocol is an algorithm, which defines how exactly to route the packet from the source to the destination. It uses one or even more of the previous metrics to evaluate the network conditions and to decide what to do. The following explores basic but important protocols for routing in wireless sensor networks. Even if they seem different from each other, they all attempt to comply with the main routing protocol requirements for WSNs:

- **Energy efficient.** Protocols need to be able to cope with node sleep and to have little overhead for route discovery and management.
- **Flexible.** Protocols must be able to cope with nodes entering or exiting the network (e.g., dead nodes or new nodes) and with changing link conditions.

5.3.1 Full-Network Broadcast

The simplest way to send a packet to any node in any network is to send the packet to all nodes in the network. This procedure is called flooding or full-network broadcast:

> **Definition 5.8.** *Flooding or full-network broadcast is defined as sending a single packet from any source in the network to all other nodes in the network, irrespective of whether they need this packet or not.*

Full-network broadcast is implemented by repeating (forwarding) the packet by each node in the network. However, if every node simply rebroadcasts any packet it receives, the network will experience a *broadcast storm*, which never ends. Thus, **broadcast** smarter techniques need to be developed, like the controlled full-network broadcast. **storm** This is illustrated in Figure 5.6 and exhibits one simple, but powerful improvement. All nodes in the network are required to repeat (rebroadcast) any packet they receive

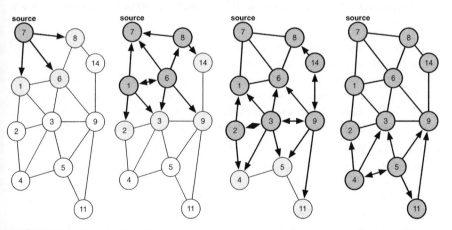

FIGURE 5.6 From left to right: The source initiates flooding by broadcasting a packet to all its neighbors (lightly shaded). In each step, the nodes which have just received the packet for the first time rebroadcast it to their neighbors, etc. Note that most of the nodes receive the packet several times, but broadcast it only once.

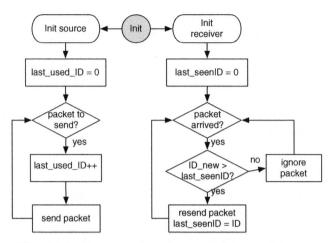

FIGURE 5.7 The general flow diagram for full network broadcast.

for the first time. Thus, every flooded packet will be repeated exactly once by each node in the network so the storm is avoided. Note that there is no guarantee that the packet will reach all nodes in the network as this depends on the topology of the network and the reliability of the links.

The only important implementation detail is how to recognize new packets from old ones. There are different strategies for this:

- Store the ID of the last seen packet. If a newly arrived packet has a higher ID, resend the packet and store the new ID.
- Store the IDs of the last X packets. Ignore all packets with IDs lower than the stored ones and resend packets with IDs that are higher or in-between the stored ones.

The general algorithm for full-network broadcast is given in Figure 5.7. Each node can serve both as a source and as a receiver.

Full-network broadcast is used to send a single message to all nodes in the network or a larger subset of them. It is simple, efficient, and quite reliable, as most nodes receive the message several times. For example, it is used to inform nodes in the network about the position of the sink or about a new software parameter.

Variations of flooding or broadcast. There are some variations of the flooding or broadcast algorithm, usually referred to as *restricted* or *limited broadcast*. Recall the flow diagram in Figure 5.7. In it, we only checked whether we already have seen this packet number before or not. In restricted broadcast, a second check is performed, which ensures that the current node also lies in some predefined region. This region has to be specified either globally in the protocol itself (not a very good idea, since too inflexible) or in each individual packet. The region can be a geographic region (specified by a location bounding box), number of hops from the original sender, or some other metric. These variations are useful in large networks, where propagating all broadcast packets to all nodes would be disastrous in terms of energy.

restricted
broadcast

5.3.2 Location-Based Routing

In this approach, the routing metric is **location**, i.e., the geographic coordinates of the nodes. The data source is aware of the geographic coordinates of the data destination. Additionally, any node of the network is aware of its own geographic coordinates and is able to ask its neighbors for their coordinates. It is not necessary to know all coordinates of all nodes or to have real geographic coordinates such as longitude and latitude. Instead, relative coordinates on a local system are sufficient, as far as all nodes are using the same system.

Location-based routing is simple. Any data forwarder (beginning at the source) compares the coordinates of the destination with the coordinates of all its neighbors and selects one to forward the data. The selection is always based on positive progress towards the destination but can also incorporate secondary metrics such as link quality.

> **Definition 5.9.** *A neighbour is defined to have a **positive progress** towards some destination, when its own distance to the sink (independently from the routing metric used) is smaller than the distance to the sink of the current node. If a routing process is always able to find a neighbour with positive progress towards the destination, it is guaranteed that the data packet will reach the destination in finite number of hops.*

Simple location-based or geographic routing is demonstrated on the left side of Figure 5.8. Here, the complete communication graph is represented with all valid

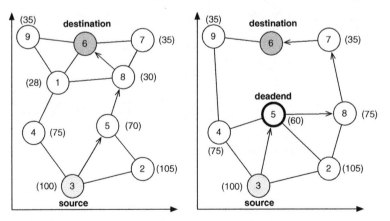

FIGURE 5.8   The left figure presents the typical scenario for location-based routing, where nodes can use the distance between the destination and their direct neighbors to select the next best hop. In this case, the routing will follow the path 3-5-8-6. The right figure instead presents the worst-case scenario, where one of the nodes (ID 5) does not have a neighbor that is closer to the destination than itself. In this case, the so called face routing will be applied to exit the deadend. The final path will be 3-5-8-7-6.

links. The routing process selects always the neighbor with minimum distance to the destination. For example, the source ID 3 selects ID 5, as it has the least geographic distance to the destination from all its neighbors with IDs 2, 4, and 5.

deadend   The main problem with geographic routing is the possibility to get into a *deadend*, which is illustrated on the right of Figure 5.8. Here, the routing process has reached ID 5, which does not have any neighbors with a positive progress towards the sink. **right-hand** In this situation, the *right-hand rule* applies. This is also called face routing mode. **rule** Generally the right-hand rule means the data packet starts traversing the network in the right-hand manner, until it again reaches a node with a positive progress towards the destination.

Remark 5.7. *The right-hand rule, and also the left-hand rule, is used to escape from a simply connected maze whose walls are connected to each other. Next time you enter a maze, put your right hand on the wall next to you and follow it. You are guaranteed to find another exit, if there is one.*

Figure 5.9 presents the flow diagram of a general geographic routing algorithm. It always selects as next hop the node with maximum positive progress towards the **face routing** sink. If it ever discovers a deadend, it initiates the *face routing* procedure.

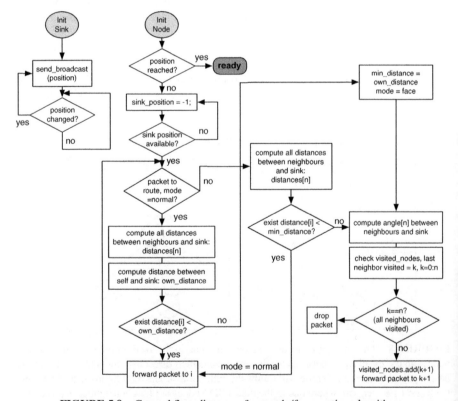

FIGURE 5.9   General flow diagram of a greedy/face routing algorithm.

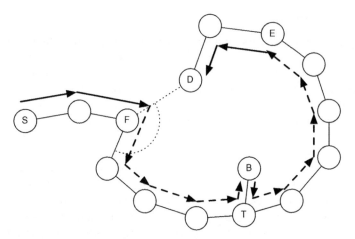

FIGURE 5.10   Greedy/face routing scenario. First, a greedy path is followed until node F, where no neighbor with a positive progress towards the destination is found. Face routing continues until node E, which finally finds a neighbor with a positive progress.

There, it first saves the distance from the face routing start node to the destination for future reference into the packet itself. Next, it selects the neighbor node with the least angle compared to the line between the current node and the destination (the right-most node). The packet is then forwarded to this node with a face start node indicator. Every next node does exactly the same, unless it gets the packet back. In this case, the next node clockwise is taken, etc. If the packet ever comes back to the face start node, it gets dropped to avoid loops in a disconnected network.

The face routing process is illustrated in Figure 5.10. From node S to node F greedy routing is followed, which takes the neighbor with the greatest positive progress towards the destination. From there, the first node clockwise is selected until reaching node T. Node T sends the packet clockwise to node B, which needs to return it to node T, as it is its only neighbor. Node T now selects the next neighbor clockwise. This continues until node E, which finally finds a neighbor with a positive progress. Greedy routing takes over again to reach destination D.

Variations of geographic routing. Geographic routing also has some variations, mainly in terms of the targeted sink position. Instead of providing the exact sink location with its coordinates, you can also provide location with radius or a region. The first is a little simpler than the second, as it requires only one location point with a radius. The result of both is the same: if the targeted region has been reached, the packet is considered delivered. Note that in this scenario, only a single node will receive the packet exactly once. This is different from a restricted broadcast, where all nodes in the targeted region will receive the packet with very high probability (recall that there is no guarantee for this, because the broadcast protocol does not take care of lost packets).

Further details about geographic routing can be found in the survey of Ruehrup [5].

5.3.3 Directed Diffusion

Directed diffusion is the name of a routing protocol, proposed by Intanangonwiwat *et al.* [2] at the beginning of the sensor networking era around 1999 to 2000. It is a very particular routing mechanism because it does not define information sources or destinations. Instead, it uses a paradigm called *information interest* and a publish-subscribe mechanism to deliver data to many different interested destinations. This routing protocol has been extremely influential in the community (the two original papers have been cited over 8,000 times) and the idea of an address-free routing protocol is especially important.

information interest

First, recall the definition of node addresses and names from Section 4.4. From this, it may seem that directed diffusion and various non-addressing routing mechanisms are focused on completely relaxing the requirement of addressing or naming. However, in reality this is only partially true. If you drop any kind of names (such as unique IDs), you will have trouble identifying loops. This can be partially solved by using a packet lifetime, but it makes the routing protocol much more vulnerable and unreliable, as the maximum lifetime needs to be tuned to the expected length of the routes. Consequently, even with address-free protocols, you will see the usage of node IDs for such purposes.

Directed diffusion original algorithm. The essence of the algorithm is the data itself. Interested destinations issue a data interest request, where they specify the type of data they want to get and how often, e.g., temperature from the region $[X_{min}, X_{max}, Y_{min}, Y_{max}]$, every second. This request is broadcasted in the whole network and each node stores the first K neighbors, from which it received the request. The assumption is that these neighbors most likely lie in the direction of the interested node and are good options for forwarding the data.

If a node identifies itself as a data source, it initiates a process called slow data delivery. This is only a fraction of the requested data delivery interval, e.g., 10 times slower. During this phase, the source sends useful data (no empty packets) to the sink by using all K identified neighbors. This also means that data starts arriving at the sink from different nodes and is heavily duplicated. However, this enables the sink to select the node, from which it received the data with the least delay. The sink reinforces this node by a special reinforcement packet. This reinforced node can now repeat the process and reinforce the neighbor, from which it received the data first, etc. After some time, the data source itself gets reinforced and a new phase called regular data delivery can begin. Now the source sends data at the requested frequency only through the nodes, from which it got the reinforcement. All phases of directed diffusion are illustrated in Figure 5.11. In the beginning of the interest dissemination phase, node 1 (the interested sink) broadcasts a message to all nodes. Each node saves at most K neighbors (in this case, K = 2) from which it first received the interest and re-broadcasts the message. At the end of this phase each node has a queue of neighbors, ordered by the time of arrival of interest.

At this point, the slow data delivery can begin, in which the source sends data over all neighbors in its queue. Once the first packet arrives at the sink, the sink reinforces the link of arrival and the reinforcement phase begins. Here, each reinforced node reinforces its own fastest link to the source. After the source is reinforced, the fast data

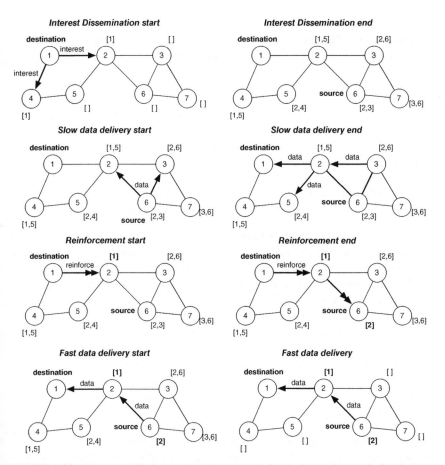

FIGURE 5.11   Directed diffusion routing phases, at the beginning and the end of each phase.

delivery phase can begin. Note that the neighbor lists get shorter for the reinforced nodes and are empty for non-reinforced nodes. This is done for optimization, as the information is not updated later in the routing process and becomes old fast.

Variations and discussion. Directed diffusion is well adapted to scenarios, where the sinks are interested in continuous data delivery. It is less adapted to situations, where only a few packets need to be routed to the destination, as the routing process requires a long "warm-up" phase.

Why the need for all four phases of directed diffusion? And, is a subset of them not enough to enable routing? Let us first consider some relevant variations of directed diffusion, before arguing when we need which version. The most commonly used variations are as follows:

- **One phase pull.** Only interest dissemination and fast data delivery phases are used. The interest dissemination phase is used to identify the fastest links to the destination and are used straight away, without first reinforcing them.

- **Push.** Interest dissemination is skipped and instead the slow data delivery starts immediately. Thus, data is being sent in an explorative manner and interested destinations reinforce the best routes directly.

The main goal of these optimizations is to save energy and minimize network traffic. However, each of these variations have their own advantages and disadvantages (see Questions and Exercises at the end of this chapter). Even though the first variation, one phase pull, makes perfect sense and typically works well in many different networks why would you ever use the full version?

The reason is that sometimes communication direction makes all the difference. In one direction, e.g., from the destination to sources, everything works fine and one particular path is fastest. In the reverse direction, the same path either fails miserably or requires much longer to arrive. There are two main reasons for this behavior:

- Link asymmetry. Refer back to the link management issues in Section 4.2.2. In these networks, the link between nodes A and B is not considered symmetric, if the communication works significantly better in one direction than in the other. Thus, a route could be working perfectly well in one direction and not at all in the reverse direction. While the interest dissemination arrives first from one particular neighbor at the source, perhaps A, the same neighbor might not be reachable from the source at all. Consequently, you need to identify the route working in the direction from the source to the destination and not vice versa and this works only in the slow delivery/reinforcement manner.
- Delay because of medium access. Recall Section 5.2.4 and Figure 5.5. The medium access protocol creates asymmetric sending delays in the network and this directly impacts the routing procedure.

Directed diffusion is an interesting and influential approach. However, it is actually rarely used in real-world applications because the idea that different sinks need different types of data is not very realistic. Most applications either process the data in-network (i.e., locally by exchanging data among neighbors only) or deliver everything to one single sink.

5.3.4   Collection Tree Protocol

The most widely used routing protocol is Collection Tree from Gnawali *et al.* [1]. It is based on the idea that data needs to be **collected** from all nodes in the sensor network rather than routed from a single source. Collection Tree Protocol (CTP) attempts to maintain a fully connected and efficient routing tree, rooted at a single sink. It is the best evaluated and tuned protocol for sensor networks currently in use. It is also a very complex protocol, so we will leave implementation details here out and leave them to the interested reader.

Chapter 4 already discussed CTP, in particular link quality estimation in terms of ETX – expected number of retransmissions (Section 4.5.5). Let us recall the most important facts. First, CTP separates data flow from control flow. That means large data packets get routed continuously, while small control packets, also called beacons are broadcasted to one-hop neighbors to update their complete route information. By keeping both flows separate, CTP attempts to minimize broadcast traffic, which is typically more expensive than unicast routing traffic. The data packets carry enough information to understand whether some neighbors have stale information, i.e., incorrect routing information. When stale information is detected, beacons can be sent to update them.

Each node in the network has a single parent, to which it forwards all data packets. Also, each node knows its current best route cost to the sink by knowing the cost to its parent (ETX) and the cost of the parent to the sink. The cost estimation for all neighbors is continuously maintained and updated, so that the node can always pick another parent.

Generally, the protocol works as follows: Once the sink boots up, it starts broadcasting a one-hop beacon, advertising its own cost, which is 0, as it is the sink itself. The beacon mechanism at any node in the network starts sending beacons once new information is available. Since no information was available before the sink booted up, all sink neighbors now start recalculating their own costs by estimating how many ETX they need to reach the sink. They achieve this by sending several beacons to the sink, which are acknowledged and thus the ETX can be computed. The beacons of the one-hop neighbors of the sinks trigger the beacon mechanism at their neighbors, and so on. All nodes exchange beacons at a high rate, until all cost estimations stabilize and each node is able to pick a parent. Once this happens, the nodes start sending data packets to their parents and they route them further to the sink. The beaconing process slows down to some minimum (e.g., a beacon every hour). An example of a stable CTP tree is provided in Figure 5.12.

What happens when something changes? For CTP, there are two types of changes: small and big ones. A node only picks a different parent if the cost of the new one is significantly lower. For CTP, this is 1.5 ETX. If a new parent is selected, the beacon process again starts at maximum and informs all neighbors about the change.

CTP is able to handle varying network properties and achieves good performance metrics in terms of delivery rate. Its cost is not negligible, though, mainly because of its broadcast beacons. Some variations and optimizations exist to handle this fact.

Variations of CTP. One of the most interesting variants of CTP is the Broadcast-Free Collection Protocol (BFC) from Puccinelli *et al.* [4]. In this, the authors reverse

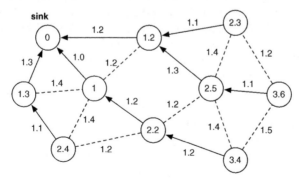

FIGURE 5.12   Example of a stable CTP tree, with node costs and link costs. Note how the cost of the nodes represents the sum of the link costs to the sink.

the direction of beacons. Instead of a sink broadcasting a beacon with its cost, each node sends opportunistically (hoping that someone will hear it) a unicast beacon to the sink. If it answers, great. If not, retry. If nothing happens, hibernate and start eavesdropping on other unicast packets until you hear about a route to the sink. Recall from Chapter 3 how unicast packets are typically handled at the MAC layer. They cannot be really tunneled to the receiver, but instead broadcasted to all neighbors until the receiver answers positively. Thus, all others sometimes get the packet too, sometimes not. This property of wireless networks is heavily used in BFC and can lower the overhead cost by as much as 70%.

5.3.5   Zigbee

It is worth exploring also a standardised routing protocol such as the one used in the Zigbee standard. There are some clear advantages to using a standardized protocol because they are often readily implemented on devices and can be used out of the box. However, they can also offer less flexibility and, in the context of Zigbee, little energy efficiency.

In general, Zigbee specifies three roles for a device in a Zigbee network:

- **Zigbee Coordinator** is what we have called the sink so far. It is the main node in the network and it is only one. It collects all of the information and also serves as security coordinator (key repository). With some variations of Zigbee, it is possible to also work without the coordinator, but having one is more usual. This node is the most powerful in the network and requires the most energy and memory resources.
- **Zigbee Router** is a device, which can sense data, but also forward data from other nodes. It needs less energy and memory than the coordinator.
- **Zigbee End Device** is a device, which can only sense phenomena and send them to its parent (a router or a coordinator). It cannot serve as a forwarder, which allows it to sleep most of the time without caring about any other devices in the network.

Each device in the Zigbee network needs to be assigned with a role, which cannot be changed during runtime. Furthermore, there are two possible modes of operation for the routers: beacon-less and beacon-based. In beacon-less Zigbee, the routers use the normal CSMA/CA MAC protocol (recall Section 3.3.3). This allows the end devices to sleep as much as they can or need, but forces the routers to stay awake continuously. With beacon-based Zigbee, the routers use a TDMA style (recall Section 3.3.2) to send out beacons to inform their neighbors when they are awake. This allows them to use a duty cycle because nodes need to be awake only when their neighbors send out beacons. The length of the duty cycle depends on the data rate of the network and is configured before deployment. However, this also requires time synchronization, which is complex as you will see in Chapter 7.

The routing protocol used is the traditional-style AODV protocol from Perkins and Royer [3]. It is based on full routing paths, where the source of the data issues a route request message, which is then broadcasted to the whole network. The route is constructed in a similar way to directed diffusion and given back to the source. Only now the source is ready to send data to the sink and starts doing so over the identified route. The biggest difference is the full routing path instead of hop-by-hop routing, which makes the protocol very unstable in case of errors or link failures. In such a situation, the routing of data packets needs to be stopped and a new route request needs to be issued.

Compared to CTP, Zigbee is less flexible and requires more configuration effort. However, it works well for small networks with little power problems such as for smart home applications. When the network grows, it becomes unmanageable to assign the roles of routers and end devices to individual nodes.

MULTI-HOP COMMUNICATIONS: SUMMARY

In this chapter, you have learned:

⇒ Routing is the process of forwarding a packet from the **source** to a **destination** via intermediate nodes called routers or forwarders.
⇒ Routing in WSNs is performed always **hop-by-hop**.
⇒ You use **routing metrics** to decide which next hop to take.
⇒ The routing metrics most often used are **hops, delay, and ETX**.

You have seen the following protocols:

⇒ **Controlled full-network broadcast** sends a single packet from one source to all other nodes of the network. It is guaranteed that it will stop once all nodes receive the packet.
⇒ **Location-based routing** selects next hops from a single source to a single destination by evaluating the geographic distance between the next hops and the final destination. Sometimes it enters a deadend, which needs to be solved via face routing.
⇒ **Directed diffusion** explores an alternative routing approach, where there are no real destinations for the packets but interests. Data is routed towards these interests and eventually arrive at the interested destination(s).
⇒ **CTP (Collection Tree Protocol)** is the most often protocol used in real-world applications. It builds a tree rooted at the sink and connects all nodes with the sink. The data gets collected from all nodes at the sink.
⇒ **Zigbee** is a (industrial) standard protocol that is less flexible than CTP.

The following are some further conclusions from this chapter:

⇒ **Routing highly depends** on the MAC and link layer protocols at least.
⇒ Routing is very **sensitive to link quality**—the more stable links are, the better the routing protocol works.
⇒ Almost none of the currently existing protocols guarantee delivery of all packets. Instead, WSN routing is **best effort**.

QUESTIONS AND EXERCISES

5.1. Compare proactive and on-demand routing in terms of their general properties (delay, costs, etc.) in the context of convergecast applications. You can assume that the convergecast application is the only application in the network. Describe a possible implementation for an on-demand and a proactive convergecast protocol. What would be the behavior of these protocols? Which one makes more sense and why?

5.2. Recall the full network broadcast protocol from Section 5.3.1. There, you were introduced to two different strategies for storing the already known packets. Answer the following questions:

    1. Why do you need such a strategy at all? What if you just resend all packets you see?

    2. Imagine that packets start arriving out of order at some nodes (e.g., first comes ID 3, then ID 2). What is the impact to both strategies?

    3. What is the impact of limited ID size on both strategies? For example, imagine you have only 2 bits for ID packet storage.

    4. What are the memory and processing requirements for both strategies?

5.3. Recall the network broadcast protocol from Section 5.3.1. There was discussion on some variations of network broadcast protocol, called restricted broadcast. Draw its flow diagram. Use a geographic limitation of the broadcasted region by providing the bounding box in every packet.

5.4. Can you design a broadcast protocol, which delivers any broadcast packets to a remote region, whose center is not the original sender? If yes, how? If not, why not?

5.5. Recall geographic routing from Section 5.3.2. There was a discussion on the variation of specifying a target region instead of a target node location. Modify the flow chart of the protocol from Figure 5.9 to implement this variation. Specify the targeted region as a location point with some radius and include this information in the packet.

5.6. Recall directed diffusion from Section 5.3.3. Draw its complete flow diagram, including all original phases. Discuss the energy requirements, the properties and the targeted scenarios for the original directed diffusion protocol and for each of its variations.

FURTHER READING

These readings provide mostly details about the previously presented routing protocols. The work of Ruehrup [5] discusses geographic routing, its mathematical foundations and various concrete geographic routing protocols. Intanagonwiwat *et al.* [2]

presents directed diffusion and its main variations. Collection Tree Protocol is presented for the first time and evaluated in the work of Gnawali *et al.* [1]. The unicast optimization of CTP, Broadcast Free Collection Protocol, is presented and evaluated in the work of Puccinelli *et al.* [4]. There exists quite a lot of other routing protocols for wireless sensor networks, but unfortunately no good survey of them. However, the work of Puccinelli *et al.* [4] offers a sound overview of collection and link estimation protocols as of 2012.

[1] O. Gnawali, R. Fonseca, K. Jamieson, D. Moss, and P. Levis. Collection tree protocol. In *Proceedings of the 7th ACM Conference on Embedded Networked Sensor Systems (SenSys)*, pages 1–14, Berkeley, California, 2009. ACM.

[2] C. Intanagonwiwat, R. Govindan, D. Estrin, J. Heidemann, and F. Silva. Directed Diffusion for Wireless Sensor Networking. *Transactions on Networking*, **11**:2–16, 2003.

[3] C. E. Perkins and E.M. Royer. Ad-hoc on-demand distance vector routing. In *Proceedings of the Second IEEE Workshop on Mobile Computing Systems and Applications*, pages 90–100, Feb 1999.

[4] D. Puccinelli, S. Giordano, M. Zuniga, and P. J. Marrón. Broadcast-free collection protocol. In *Proceedings of the 10th ACM Conference on Embedded Network Sensor Systems*, SenSys '12, pages 29–42, New York, NY, USA, 2012. ACM. ISBN 978-1-4503-1169-4.

[5] S. Rührup. Theory and Practice of Geographic Routing. In Hai Liu, Xiaowen Chu, and Yiu-Wing Leung, editors, *Ad Hoc and Sensor Wireless Networks: Architectures, Algorithms and Protocols*. Bentham Science, 2009.

6

DATA AGGREGATION AND CLUSTERING

This chapter explores how to manage very large wireless sensor networks. In these networks, typical data collection techniques are not viable, as they flood the network with too much data and exhaust its energy reserves too quickly. There are several possibilities to minimize the traffic in very large WSNs. You can aggregate the data on its way to reduce the number of packets; organize the network into sub-networks and let the sub-networks decide what to do with the data and, last but not least, reduce the data produced at the individual nodes in the first place. This chapter discusses all of these techniques and their properties.

If you are not interested in large networks or you are already proficient in this topic, you might still want to have a look at Section 6.3, which explores Compressive Sensing, a rarely known technique that is also useful for small networks, producing large data (so called big data).

6.1 CLUSTERING TECHNIQUES

Clustering was one of the first techniques applied to very large networks. The main idea is to organize the network into smaller sub-networks, so that data can be collected and analyzed in a location-restricted way, meaning only some important or aggregated data comes to the final network-wide sink. The idea is depicted in

Introduction To Wireless Sensor Networks, First Edition. Anna Förster.
© 2016 The Institute of Electrical and Electronics Engineers, Inc. Published 2016 by John Wiley & Sons, Inc.

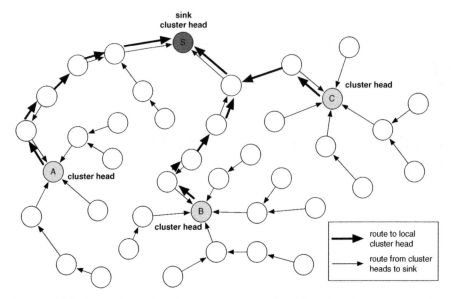

FIGURE 6.1   Example of clustering a large sensor network into four clusters. The cluster heads represent local mini-sinks and cluster head S is simultaneously the global sink of the network. The routes from individual nodes to cluster heads are represented, as well as the routes from cluster heads to the global sink.

Figure 6.1. In clustering, we differentiate between the following roles of individual nodes:

- **Cluster members** are simple nodes, which sense some phenomena. They send their data to their cluster heads. A cluster member belongs to a single cluster head.
- **Cluster heads** are local sinks, which gather all of the information of their cluster members. Several options are possible here and depend on the application: send everything through a more powerful link to a final data storage place; aggregate or compress the data then send it through the sensor network to a global sink; or analyze the data and either report some event or not.
- **The sink**, sometimes called the global sink, is a single node in the network, which gathers all information from all nodes in the network. It is not necessary that it exists at all.

What is the difference between clustering a large network and deploying several small ones? When deploying several small networks, they will be independent of each other and at the same time the engineer needs to make sure that they do not interfere with each other. This approach is rather cumbersome and inflexible. In a large network, you can take advantage of all nodes and allow the nodes to cooperate to achieve their goal.

The challenge is that the network is too big to be handled by normal data collection protocols.

The main objective of clustering is to save energy and to avoid data congestion in the network. The underlying assumption is that if you try to route all of the data of all nodes to a single sink, then at least some of the nodes will be completely congested (the ones close to the sink) and that their batteries will be drained too fast. Furthermore, it is important which are the cluster heads and how they are selected and managed. Some concrete examples are discussed next.

6.1.1   Random Clustering

The simplest approach to clustering is to randomly select cluster heads from all available nodes. A protocol like this is LEACH by Heinzelman *et al.* [4] and was the first clustering protocol. In fact, it was the first routing protocol as well. Before LEACH, there existed the assumption and the concept that all sensor nodes communicate directly with the sink in a single hop. This was the beginning of wireless sensor networks research and at the time, the main new concept was tiny size and resources.

LEACH works as follows: Every node throws a dice and decides whether to become a cluster head or not with some probability ϵ. If it becomes a cluster head, it advertises itself to all nodes around. Non-cluster head nodes decide which cluster head to join by selecting the best communication link, which also offers the possibility to select a lower transmission power. Nodes start sending their data to their cluster heads (one hop directly) and cluster heads send it to the sink (again one hop). In order to not drain the batteries of the cluster heads too fast, cluster heads change over time. Thus, the procedure is repeated regularly.

The advantages of LEACH and its subsequent protocols and optimizations is that it is simple and flexible. It can be easily extended to also work with fixed transmission power and several hops to the cluster heads or from cluster heads to the sink. However, it also has several disadvantages. First of all, the random selection of cluster heads cannot guarantee any uniform distribution of the cluster heads in the field. Thus, some areas will have many cluster heads but some will have none. This results in a very unbalanced transmission costs in the network (Figure 6.2).

Another problem comes from the fact that in really large networks, even with maximum transmission power, the nodes cannot reach all other nodes.

Thus, the original version of the algorithm might leave some areas without cluster heads at all and also without data reporting for long periods of time. A remedy was introduced, where a node throws the dice again, if it does not hear about a cluster head. However, this results in too many cluster heads and again in higher energy expenditure.

Another disadvantage of the approach is its "jumpy" behavior. To save energy on the cluster heads, the protocol selects new ones. As discussed, you need the multi-hop version to ensure that all nodes are connected to the global sink through cluster heads. Thus, every time the cluster head changes, you need to restart the routing procedure, as all existing routes become unusable.

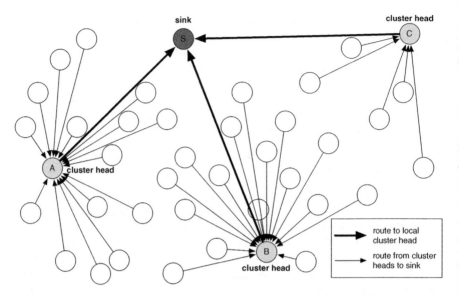

FIGURE 6.2   Example of random selection of cluster heads with LEACH. While two of the clusters are large, one (cluster C) is small.

6.1.2   Nearest Sink

Another possibility to tackle communication problems in large networks is to install more than one sink. This means more than one node in the network serves as a gateway to the final storage place. Imagine that all cluster heads in Figure 6.1 have an upload link to the database and are thus preselected as cluster heads. Exactly as for the global sink, these multiple sinks have bigger batteries and thus less energy problems.

The task of the clustering algorithm is how to select the cluster head for each individual node. The solution is relatively simple. Each cluster head announces itself as a sink, similar to how you found the sink in routing algorithms in the previous Chapter 5. This information is updated as normal routing information at every hop (e.g. with ETX count, hop count, etc.). In other words, you run the sink discovery procedure for each of the sinks individually. Every node receives several sink announcements from various neighbors. It simply selects the sink which looks best (depending on the used routing metric) and starts using that one. It can also keep a secondary choice, in case something happens with its first choice.

This implementation has the disadvantage that the cluster heads are actually fixed and need some special hardware. This renders the network more expensive than using only normal nodes. It can be also seen as several networks installed next to each other. On the one hand, as the beginning of this chapter noted, this increases the installation and maintenance costs and lowers the flexibility of the network. On the other hand, this is the preferred solution by practitioners. It offers redundancy for the sink (if one fails, the others can take over its duties, even if at higher communication costs). Overall, it is simple, stable, and well balanced.

6.1.3   Geographic Clustering

The preceding examples show that clusters are geographically organised, meaning one cluster is built by geographically close nodes. This leads to the idea that you can also use direct geographic information to build the clusters. An example of such a protocol is Clique [2]. The protocol itself is quite complex, as it relies also on reinforcement learning to select the best cluster head in geographic clusters. Here, the discussion presents a simplified version of it.

First, the network is cut into clusters according to geographic coordinates. Figure 6.3 presents a sample network, where the cluster size is set to be 100×100 meters. The remaining challenge is to select the cluster heads inside the geographic clusters. You could, for example, select them randomly. Exactly as with LEACH, each node throws a dice and decides whether to become a cluster head or not. If a node does not hear about a cluster head in its own cluster, it retries the procedure.

Another possibility is to exchange the IDs of all nodes in the cluster and to select the lowest ID as the first cluster head, then the second-lowest as the next cluster head, etc. Also the geographic location can be used as a criteria for selecting the cluster head, e.g., the closest to the sink.

The problem of selecting a cluster head in a geographically restricted area is simpler than in a large network. Both random selection and deterministic selection work well, as the communication area is small and there is no need to balance the energy. This renders geographic clustering an efficient way to cluster a large network into smaller, independent pieces.

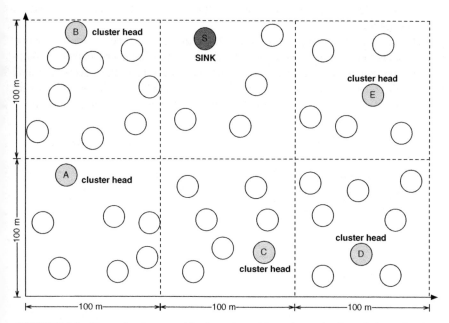

FIGURE 6.3   Example of geographic clustering with random selection of cluster heads.

6.1.4   Clustering Summary

All of the previous clustering techniques tackle the problem of large networks at the level of organizing and splitting the communication flow. However, there is also the possibility to process the data on the way to the sink (combined with clustering or not) and thus to reduce the amount of data incoming to the sink. Section 6.2 explores this concept.

6.2   IN-NETWORK PROCESSING AND DATA AGGREGATION

There are mainly two concepts for in-network aggregation and processing: compression and aggregation. In compression, data remains as it is, but its resource usage is minimized. For example, 5 packets with 5 bytes data in each could be easily combined in a single packet and this would save the communication overhead of sending 5 different packets. In aggregation, the data is somehow processed and only part of it continues to the sink. For example, only the maximum sensed value is allowed to travel to the sink and all lower values get discarded on their way. The following sections discuss some examples.

6.2.1   Compression

The first choice for compression is combining data from several packets into a single one. This concept can always be used and delivers very good results. However, it also relies more or less on the arrival of packets at the sink and thus cannot be used for real-time applications.

Let us explore the example in Figure 6.4. On the left, you see a sample network, where data is routed with the help of the CTP routing protocol (see Chapter 5). Without compression, nodes 4, 7, and 10 would need to forward 2 packets every sampling period—their own packet and one from a child node. Node 5 would need to forward 3 packets and node 3 even 4 packets. This sums up to a total of 17 transmissions. If

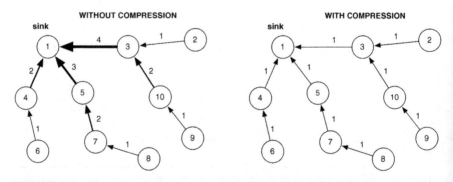

FIGURE 6.4 Comparison between a network without (left) and with compression (right) of individual packets implemented. The number of transmitted packets are given next to the individual links.

we simply combine the packets at each hop into a single one, each node would have to transfer only 1 packet, which sums up to only 9 transmissions, exactly as many as nodes exist in the network. Recall that transmissions are costly because each of them needs time to obtain the wireless medium and require not only the data to be transferred, but also various protocol headers, control packets, acknowledgements, etc. By combining the data into single packets, you would save 47% of the transmissions.

The only tricky part of this measure, which is otherwise beautifully simple, is how to synchronize the data packets so that data packets are not delayed too much. The problem occurs most often when clocks of individual nodes start to drift away from each other. Probably the most trivial method to implement this is to define a waiting period exactly as long as the sampling period. Thus, every node sends a single packet exactly after it has sampled its own sensors. Everything, which has arrived between its last sampling period and now, gets into the same single packet and out. This means that in the worst case scenario, packets from the previous sampling round from other nodes will be forwarded only in next sampling round. This delay might be large, but it is bounded so it is the maximum possible delay.

Huffman Codes   There is also another potential problem. Sometimes the data is so large that several pieces of it do not fit into a single packet. Then, you need to introduce some real compression techniques to minimize its size and to fit more of it into a single packet. Here, data compression algorithms come into play such as Huffman codes. A Huffman code assigns a special code to each character, which is shorter than its original encoding. Let us explore an example.

Assume that you have an alphabet with 5 characters only. To encode those with bits, you need at least 3 bits. The codes for these characters could be 000, 001, 010, 011, and 100. The Huffman code calculates first the probability of how often individual characters occur. For example, after analyzing some sample text in this alphabet, you can identify the probabilities for your characters as 0.1, 0.3, 0.2, 0.15, and 0.25. This is depicted in Figure 6.5. Now, you can create the Hufmann tree. For this, the

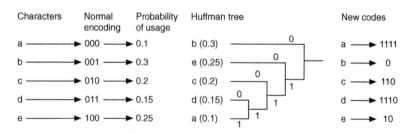

FIGURE 6.5   An example of creating a Huffman code for an imaginary alphabet of 5 characters.

characters are ordered by their probability and then the two lowest ones are connected with a next level node, until the root is reached. Now, you can start assigning codes to the individual nodes. Assign a 0 to the node with the character b (the one with the highest probability) and to everything else a 1. Next, assign a 0 and 1 to the next level of nodes and you get 10 for the character e and 11 for everything else, and so on.

The codes are now clear. However, how do you decode them? If you look at the encoded example from Figure 6.5, you do not have empty spaces between the characters (they are introduced in the figure just for better human readability). Instead, you have the continuous code of 1111001101110000101010. So, if you start with the first character, you get a 1. It cannot be a b, because b has the code of 0. Thus, it is something else. If you continue, the code is now 11. This still does not correspond to any of your codes, so you just continue like this until you get something you know. In this case, the code 1111, which is an a. Now you are ready with the first character so you can continue on.

The real beauty of the Huffman code lies in the used probabilities to create codes. If you look carefully at the created tree, you see that the characters used most (with the highest probability) get the shortest codes, like b and e. The characters used rarely, like a and d, get the longest codes. Thus, this code is optimal for this alphabet (and the text used for calculating the probabilities).

Of course, the Huffman code is only one out of many different codes which **encode GPS** you could use in sensor networks. One favorite trick to apply is to *encode GPS* coordinates. GPS coordinates typically occupy 40 bytes in total for the longitude (8 bytes), latitude (8 bytes), time (16 bytes), and elevation (8 bytes). Instead of storing the data as floating numbers, you could store it as ASCII characters. Here is an example:

```
59.271327 17.846174 2008-10-14T16:24:35Z 1.375
```

These are 49 characters, 1 byte each, which makes 46 bytes. This is, for now, more than you had before. Provided that a sensor network is rather location-restricted, you could first eliminate some of the data by cutting away the leading numbers. For example, you can cut the 59 in front of the latitude and the 17 in front of the latitude. Including the commas, which you also do not need any more, you have 6 bytes less. This is already as large as your original storage. Now, you are left only with numbers and several special characters, a total of approximately 15 characters. These can be easily encoded with a total of 4 bits each (even without Huffman codes), which will result in a total of 20 bytes. This is half of what you had before. Applying Huffman codes will also help lower this number further. This calculation does not even consider the fact that some of the leading numbers of the time can be eliminated or that the elevation is typically so error-prone, that you can cut it all together.

Differential Compression and Model Prediction Cutting unnecessary data leads to the topic of differential compression, which extracts the new data compared to something already existing. This technique is widely used for transferring large files, synchronization services such as remote storage, etc. In sensor networks, it is simpler. A node can transfer first all of its available data to the sink, e.g., its co-ordinates, current temperature, current humidity. Then, at every sampling period, it normally samples its sensors, but it only reports to the sink if something, maybe even significantly, changed. For example, a common concept is to have one sampling period (e.g., 1 minute) and one minimum reporting period (e.g., 30 minutes). Then, the node reports only if the sampled data is different than the last reported one or when the minimum reporting period expires. In this way, the sink receives only few data items, but it knows that the node is still alive. This type of compression is often used for sensor networks such as to disseminate new application code to all nodes, as in the work of Rickenbach and Wattenhofer [5].

Model prediction is very similar to differential compression, but uses a more general concept of what "same data" means. Here, a model is built for the data, which mirrors its normal fluctuations and changes. For example, a model of temperature data in a house mirrors the fact that it is normal for the temperature to fluctuate between 21 and 23°C, with small steps of 0.5°C over the span of 5 minutes. If the measured data differs from the model, the data needs to be sent to the sinks, e.g., when the temperature rise is too fast or when the minimum or maximum values are exceeded. A protocol that uses such models is the Spanish Inquisition Protocol from Goldsmith and Brusey [3].

Compression Summary This discussion shows that compression in sensor networks requires the following:

- Good understanding of the data. This enables more memory-efficient storage and elimination of repeated data.
- Correct formatting. The normally used formatting, such as floating point numbers for GPS coordinates, is typically the worst choice. Instead, smarter and more efficient formats can be found, which also allow for compression.
- Simple compression. A sensor node cannot afford a complex compression algorithm and it is usually not needed (unless for video or audio streams, etc.).

Data compression techniques with their underlying information theory concepts (such as the probability analysis of characters in Huffman codes) is extensively studied for example in the book of Sayood [7].

6.2.2 Statistical Techniques

Compression is very useful and should be considered first when communication in the network is heavy. However, there are also other techniques. One of them is to evaluate the data in a statistical way somewhere on the way from its source to the

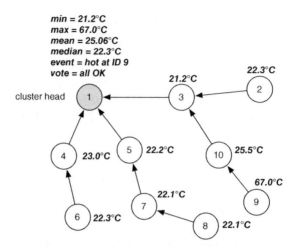

FIGURE 6.6 An example of cluster-level aggregation of temperature data. Various statistics can be computer at the cluster head and then transferred to the sink.

destination. This can be done on the cluster head, which also offers the advantage of having a location-restricted data set. Consequently, instead of sending all the data to the sink, only the newly calculated values are sent, which are typically much smaller than the original data. This can also be performed the whole way to the sink, e.g., on each hop.

voting   Statistical evaluation could be the computation of some statistical values of the complete data set such as minimum, maximum, and mean. It can also be a *voting* consensus   mechanism, in which the median of the data set is calculated. Or it can be a *consensus* approach, in which individual nodes vote whether some event has occurred or not. While such an evaluation might seem very practical and efficient at first glance, it needs to be carefully evaluated before usage.

> Once statistically evaluated, the original data cannot be recovered any more.

In Figure 6.6, a single cluster of nodes with their cluster head is shown. The temperature is sensed at each node and transmitted to the cluster head. The cluster head can calculate various statistics over the data such as the minimum, maximum, mean, and median temperatures sensed. It can also report only events, e.g., the fact that it is really hot at node 9. Or it can define a voting algorithm in which at least two nodes have to report a "hot" or "burn" event. In all these cases, the original data is lost and cannot be recovered at the sink any more. What does the "hot" event at node 9 mean? Perhaps it is faulty? Or maybe the other nodes do not report a "hot" event because they are only slightly below the threshold for it? Perhaps node 9 is almost burning already? If too many statistics are transferred to the sink, there will be no real communication savings. If too few statistics are transferred, the question remains whether you can use the data at all.

However, this does not mean that in-network aggregation is never good. There are certain applications, which precisely define their events and what "normal" behavior is. For example, in structural monitoring mechanical stress is measured such as acceleration (vibration). High-frequency data like acceleration is gathered at the nodes and needs to be processed and compared to various thresholds.

Then, only the final event is transferred or, if the event is considered important, the full data can be sent to the sink as an exception. In such high-frequency data domains, you have only two options: either drastically reduce the data volume after sensing or sense less. Section 6.3 discusses this second option.

6.3   COMPRESSIVE SAMPLING

Instead of first sensing a large amount of data then trying to reduce it, it is much better to sense only as much as you can transfer. However, how do you know when something interesting is happening such as an earthquake? This is the task of compressive sensing—to decide when to sense data so that nothing interesting is missed. It is not only smarter to do it this way but sometimes, it is simply impossible to fit all of the data on the sensor node or the sensing drains too much power on the node.

Compressive sampling or sensing is a research topic within the area of signal processing. This discussion does not enter into mathematical details of compression sampling but the interested reader can turn to the tutorial of Candes and Wakin [1]. The Randomized Timing Vector (RTV) algorithm by Rubin and Camp [6], which uses compressive sensing, is easier to understand and follow. The main idea is that instead of continuously sensing a signal, such as an ECG signal or vibration signal at some base frequency, you can sample it only from time to time and still preserve the information underlying the signal. Most signals carry some information and some noise. You are actually interested only in the information. Thus, you want to take as many measurements as needed to get the information but ignore the noise. An example is given in Fig. 6.7. A signal (a repeated pulse) is compressively sensed (the red dots) while its information properties are preserved. In this case, the information is the repeated pulse itself with its frequency and strength.

RTV works as follows: It first decides on a base sampling frequency, e.g., 500 Hz (once every 2 milliseconds). Then, it decides on a sampling period, e.g., 1 second. Then, it creates a random vector with an exact length of as many samples that can fit into a sampling period. In this case, you need 500 samples. Each of these numbers is either 0 or 1. Now, a timer at the base frequency is initialized and every time it fires, the application looks at the next number in the random vector. If it is a 1, then it takes the sample. If it is a 0, it goes back to sleep. Once the sampling period is over, the random vector is recalculated and the resulting samples' vector is sent to the sink. This vector is much smaller than the full sample vector and its length depends on how many 1s you have in the random vector.

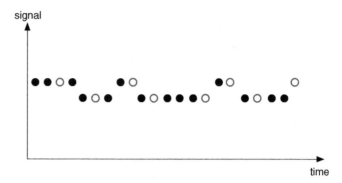

FIGURE 6.7   An example of compressive sensing over a signal. Since the signal has some well-defined properties (in this case, repeatability), it can be compressively sensed without information loss. The original signal consists of all dots and the compressively sensed ones are the circles.

At the sink, the sampled vector can be analyzed and research has shown that the underlying information is preserved. Thus, with this type of compression you cannot directly reconstruct the original signal, but its properties are preserved and the resulting smaller signal can instead be analyzed.

Compressive sampling is a powerful approach to reduce the amount of produced data at its source. It is quite similar to compression techniques but works before the data has been sampled.

DATA AGGREGATION AND CLUSTERING: SUMMARY

This chapter discussed four main concepts to minimize communications in very large sensor networks:

⇒ **Clustering** organizes the network into sub-networks called clusters. Clusters can be identified randomly or geographically. Each cluster gathers all of its data on a single node, called the cluster head, and then sends the data in a compressed or processed way to the sink.

⇒ **Compression** refers to minimizing the required communication overhead and storage for individually sampled data. For example, several data items can be unified in a single data packet or data compression techniques can be applied to further reduce their volume.

⇒ **In-network processing** refers to calculating some statistical values or to allow voting between the nodes of a cluster. For example, only the mean values can be transferred to the sink.

⇒ **Compressive sensing** refers to a concept from signal processing, which randomly samples a signal instead of continuously. This reduces the sampled data at the source directly and no further compression is needed.

It is important to note that:

⇒ Clustering is a communication organization measure and needs to be used together with other techniques to achieve communication savings.
⇒ Compression and compressive sensing preserve the full properties of the originally sensed signal, and sometimes it can even be completely reconstructed. These two concepts are our preferred options.
⇒ Statistical in-network processing does not allow for reconstruction of the original signal, nor of all of its properties, and has to be used very carefully.

QUESTIONS AND EXERCISES

6.1. Recall the first option for data compression from Section 6.2.1. For Figure 6.4, we calculated that we can save 47% in the number of required transmissions by simply combining the data from individual packets into single packets. Try to also calculate the savings in number of bits transferred. Assume that each individual packet (as created at each individual node) consists of 20 bytes of data and 20 bytes of overhead (control information, header, etc.). How many bytes will be saved totally in the network for one sampling round? You can assume that only data packets are sent, no control packets.

6.2. Next, assume also that control packets are sent. Two control beacons are needed for each data packet, 10 bytes each. How many bytes will you save now totally for the network?

6.3. Next, introduce ETX for the individual links. You can randomly select some numbers between 1 and 3 for the individual links (do not forget to write them down for your solution). Calculate how many bytes of data will be sent in this network in the case of no compression and with compression. Now you have a pretty good idea of how much savings you can achieve using this simple measure.

FURTHER READING

LEACH was first presented in the work of Heinzelman *et al.* [4]. More information about the geographic clustering approach called Clique can be found in [2]. A very good introduction to general data compression techniques can be found in the book of Sayood [7]. Model-based evaluation of sampled data and selective sending to the sink using the Spanish Inquisition Protocol is presented and evaluated in the work of Goldsmith and Brusey [3]. The same techniques applied to source code dissemination is used in the work of Rickenbach and Wattenhofer [5]. Finally, compressive sampling or sensing is well described in the tutorial of Candes and Wakin [1] and applied to sensor networks in the Randomised Timing Vector protocol for vibration data in Rubin and Camp [6].

[1] E. J. Candes and M. B. Wakin. An Introduction To Compressive Sampling. *Signal Processing Magazine, IEEE*, 25(2):21–30, March 2008.

[2] A. Förster and A. L. Murphy. CLIQUE: Role-Free Clustering with Q-Learning for Wireless Sensor Networks. In *Proceedings of the 29th International Conference on Distributed Computing Systems (ICDCS)*, Montreal, Canada, 2009.

[3] D. Goldsmith and J. Brusey. The spanish inquisition protocol: Model based transmission reduction for wireless sensor networks. In *Sensors, 2010 IEEE*, pages 2043–2048, Nov 2010.

[4] W. Rabiner-Heinzelman, A. Chandrakasan, and H. Balakrishnan. Energy-Efficient Communication Protocol for Wireless Microsensor Networks. In *Proceedings of the 33rd Hawaii International Conference on System Sciences (HICSS)*, page 10pp., Hawaii, USA, 2000.

[5] P. Rickenbach and R. Wattenhofer. Decoding Code on a Sensor Node. In *Proceedings of the 4th IEEE International Conference on Distributed Computing in Sensor Systems*, DCOSS '08, pages 400–414, Berlin, Heidelberg, 2008. Springer-Verlag.

[6] M. J. Rubin and T. Camp. On-mote compressive sampling to reduce power consumption for wireless sensors. In *Sensor, Mesh and Ad Hoc Communications and Networks (SECON), 2013 10th Annual IEEE Communications Society Conference on*, pages 291–299, June 2013.

[7] Khalid Sayood. *Introduction to Data Compression*. Morgan Kaufmann, 4th edition, 2012.

7

TIME SYNCHRONIZATION

This chapter focuses on a service called time synchronization, which is often taken for granted in modern computer systems. The resource-limited design of wireless sensor networks makes it unviable to always require in-system time synchronization. The application developer has to make sure that he or she has implemented the required time precision. The precision can highly fluctuate, from minutes for high-end latency-tolerant monitoring applications to milliseconds for medium access control for high-throughput networks. This chapter explores the challenges and requirements for implementing time synchronization as well as the basic techniques, especially for low-precision protocols.

If you have already implemented other time synchronization protocols, you are advised to jump directly to this chapter's Summary. If you believe you will not need time synchronization in your application, do not skip this chapter, sooner than later you will need it.

7.1 CLOCKS AND DELAY SOURCES

A clock is a device, which ticks regularly and counts those ticks. In wall clocks, there is the pendulum, which takes over the ticking. This "ticker" is also called the **timekeeping element**. Every x-th tick, one of the hands of the clock is moved one step forward.

In microcontrollers, the clock is not very different. The processor has a fixed frequency, which is used as the timekeeping element. There is also the possibility to use

Introduction To Wireless Sensor Networks, First Edition. Anna Förster.
© 2016 The Institute of Electrical and Electronics Engineers, Inc. Published 2016 by John Wiley & Sons, Inc.

an external quartz to take over this role, but this is more expensive. A simple register is used to count the ticks. Instead of really counting them, its value is simply incremented at every tick. Once it overflows, another counter is increased, which is the actual time register.

Before exploring the properties of clocks and why time synchronization is needed, here are some basic definitions.

> **Definition 7.1.** *Clock resolution or rate is the step at which the clock progresses and can be read.*

> **Definition 7.2.** *Clock skew or drift is the difference between the progressing speeds of two clocks.*

The clock skew is the main reason why you need time synchronization. It makes two clocks drift from each other and this causes problems, as the next Section 7.2 discusses.

> **Definition 7.3.** *Clock offset is the difference between the time shown by two different clocks.*

> **Definition 7.4.** *Clock jitter is when the individual ticks of a clock are not perfectly the same.*

Figure 7.1 graphically presents the above properties of clocks. Next we explore what are the problems when time is not correct.

7.2   REQUIREMENTS AND CHALLENGES

Time is something very natural for people but it is not natural for communication networks. Recall from previous chapters how often you have used the notion of timers to do something regularly or the notion of synchronous activities such as sleeping or waking up the sensor nodes. What happens when time is not perfectly the same at individual nodes, but drifts apart?

A time synchronization protocol corrects the current local time to match a shared time. In other words, it overwrites the currently available timestamp with a newer, correct one. However, the process is rather complex.

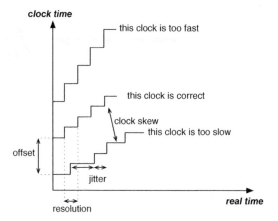

FIGURE 7.1   The properties of clocks.

Two clocks drift apart even after synchronization. This is the main reason why time synchronization protocols **regularly** need to synchronize the clock and cannot do it only once.

The maximum drift rate of a clock dictates the synchronization period. The maximum drift rate ρ is provided by the manufacturer and is expressed in ppm (parts per million). For example, if you know that $\rho = 20$ ppm for a particular clock, you know that this clock differs from reality at the maximum rate of 20 seconds per one million seconds (the seconds can be exchanged with any other time metric). This is a drift of 1 second per approximately 13.8 hours. This also means that two identical clocks will drift from each other with at most 2ρ after synchronization. If you want to limit the drift to δ seconds, you need to resynchronize the clocks every τ_{sync}:

$$\tau_{sync} = \frac{\delta}{2\rho} \qquad (7.1)$$

The time synchronization protocol cannot jump back and forth in time. Correcting the timestamp in any direction can have some nasty consequences. For example, events can be skipped or events can be repeated. Thus, time correction is allowed to progress only in one direction – into the future.

A time synchronization protocol can be evaluated with two main performance metrics, accuracy and precision.

Definition 7.5. *Time synchronization accuracy is the maximum offset between the synchronised clock and a reference clock (for example global time).*

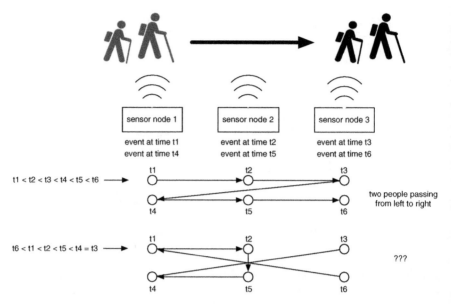

FIGURE 7.2 An example of time ordered events from a sensor network. When the events are correctly timestamped, two people can be correctly detected walking from left to right. When the events are incorrectly timestamped, no meaningful conclusion can be done.

> **Definition 7.6.** *Time synchronization precision is the maximum offset between any two synchronised clocks.*

Time synchronization precision is important to understand the sensor data.
The reason why time synchronization is important was already discussed for communications and for correctly running applications at the sensor nodes. However, it is even more important for understanding the sensory data arriving from the network.

Figure 7.2 presents an example, in which movement sensors monitor a hiking path. When a movement is detected, the node timestamps the event and sends it to the sink.

Three sensors are used to monitor this part of the hiking path and the sink receives events from all three of them. The time order of these events reveals the direction of movement at the hiking path, which is as important as the event of the movement itself. Furthermore, it enables the sink to count the number of people passing through. Figure 7.3 shows the correct event ordering for two people passing and an incorrect version, resulting from incorrect timestamps. It becomes obvious that the requirements of the applications need to be used to calculate the required time synchronization precision.

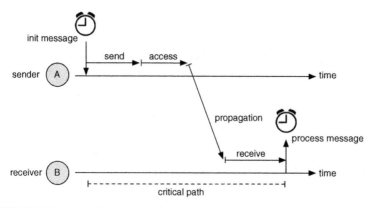

FIGURE 7.3   A sender-receiver synchronization process with all sources of delay.

A clock can be synchronized either internally or externally. An *external synchronization* is a synchronization with the support of an external time source to which all other clocks are synchronized. An *internal synchronization* avoids the usage of an external clock source and instead targets a consistent view of time across all synchronized clocks. Externally synchronized clocks are always also internally synchronized but not vice versa.

external synchronization

internal synchronization

7.3   TIME SYNCHRONIZATION PROTOCOLS

The following sections discuss concrete time synchronization protocols for sensor networks.

7.3.1   Lightweight Tree Synchronization

The lightweight tree synchronization protocol (LTS) by van Greunen and Rabaye [3] uses sender-receiver synchronization, in which a sender provides the receiver with the correct time and the receiver calculates its offset so it can correct its own time to match the time of the sender. Figure 7.3 presents the main idea and the individual steps of the process. It is important to understand that there will always be some residual offset, which you cannot really accommodate. For example, when the synchronization message is created and timestamped, the node still needs to send it to radio and the radio needs to access the wireless medium. The time spent in such functions is called the *critical path*. This delay is not accounted for in the offset calculation later, but it can be inserted as an experimentally obtained mean value.

critical path

LTS itself works as follows: It first builds a spanning tree of the network, rooted at the sink, which has real-time available. The root then sends messages to its children and synchronizes them. The synchronized nodes synchronize their children and so on. The synchronization precision clearly depends on the depth of the tree and thus the protocol does not scale well for large multi-hop networks.

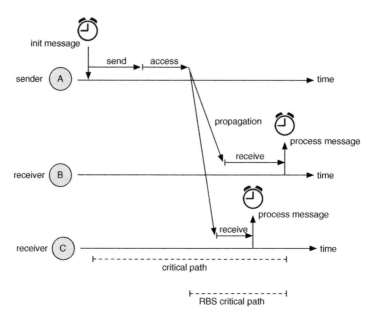

FIGURE 7.4   A receiver-receiver synchronization process. Its critical path is shorter than for sender-receiver synchronization.

7.3.2   Reference Broadcast Synchronization

The main disadvantage of the LTS protocol is the used sender-receiver synchronization process because the critical path between initiating the message at the sender and processing it at the receiver is long. This can be avoided when the receiver-receiver synchronization is used, as shown in Figure 7.4. The scenario is the following: A sender normally initiates a synchronization message and sends it out via broadcast to all its neighbors. The neighbors are all receivers of the message and receive it at approximately the same time. This means they all experience exactly the same delay caused by the sender and that the remaining indeterministic part is smaller, resulting from the propagation delay and their own receiver functions. Thus, the critical path is shorter and the incurred delay lower. Both receivers from Figure 7.4 can now exchange the information they have received and calculate their offsets more precisely.

Reference broadcast protocol (RBS) [1] makes use of this receiver-receiver synchronization process. It simply broadcasts messages across all nodes in the network. With increasing number of broadcasts, the mean overall offsets can be used and the total precision increases. RBS scales well with large networks but is not very efficient in terms of communication overhead.

7.3.3   NoTime Protocol

Strictly speaking, this last protocol does not synchronize the time at the nodes at all. Many applications do not really need a synchronized time directly on the

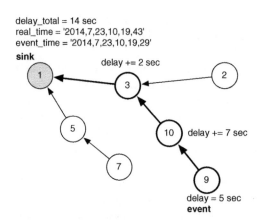

delay_total = 14 sec
real_time = '2014,7,23,10,19,43'
event_time = '2014,7,23,10,19,29'

FIGURE 7.5   A multi-hop network with NoTime synchronization protocol running on the nodes. Each node updates a field in each data packet that it forwards with the delay this packet has experienced at this node. The sink can then calculate the real time from this delay.

sensor nodes. For example, a simple network with a single sink and no time-dependent communication protocols only needs timestamps for the sensor events delivered to the sink. In these networks, the problem of synchronization can be inverted. Instead of delivering time data to individual sensor nodes and thus synchronizing them with the sink's time, the sensor nodes deliver all necessary information to the sink, so that the sink itself can calculate the real time of the sensor events. This is called the NoTime protocol and it is a widely used algorithm, although rarely described in detail or systematically evaluated.

Figure 7.5 presents a sample multi-hop network with several nodes. Each of them runs its own local time counter. When a sensor event occurs, it is not timestamped, but instead gets a field called delay. The creator node of this event calculates how much time passes between the creation of this event and its sending out to the next hop. This delay can be extremely varied, ranging from a few milliseconds to seconds, and depends on the hardware and on the communication protocols used. This final delay is written into the data packet just before sending it out to the next hop.

The next hop does almost the same—it calculates how much time the packet has spent at this node before being sent out again. This delay is added to the existing one in the packet.

In Figure 7.5, node 9 creates a sensor event and computes the delay to be 5 seconds. The next hop, node 10, adds another 7 seconds to the delay. The next hop, node 3, adds another 2 seconds. The sink is the next hop and receives a total delay of 14 seconds. The sink is the only node in the network which has the real time, since it is connected to the outside world. Thus, the sink can easily subtract the received delay from the current real time and get the real time of the event. This can now be written into the event data and stored.

The main advantage of this protocol is that you spend absolutely nothing for time synchronization. Instead of running an extra protocol and its overhead, you in fact save resources. This is due to the fact that you do not send real timestamps, which are at least 32 bits long, but only 8 to 16 bits for the delay, depending on the required precision. The precision of the protocol depends on three sources of delay:

- **Propagation delay.** We have assumed that the actual communication between two nodes does not incur any delay. In normal sensor networks, where nodes are only tens of meters apart from each other at most, this is an acceptable assumption. The propagation delay over 100 meters, for example, is only approximately 5 μs. However, if the network has long links, the propagation delay might be significant, such as 50 μs for 1 km. Over many hops, this delay adds up and greatly impacts the precision of the protocol.

- **Hardware delay.** Another delay, which was not accounted for, is the hardware delay on the node to perform needed operations such as writing the delay into the packet before sending it out. A careful implementation should be able to minimize this delay to an insignificant level.

- **Local time resolution.** If the node can only differentiate between seconds, you cannot expect the sink to compute the real time with precision in microseconds. In fact, this rounding up at each node of the supported resolution adds up over multiple hops.

Overall, the NoTime protocol's precision depends mostly on the number of hops the data packet travelled. It is a great protocol to use when the application is not time sensitive and the needed precision is not high (e.g., in the range of a few seconds).

TIME SYNCHRONIZATION: SUMMARY

Time synchronization is extremely important for sensor networks both for the communication protocols used and for the needs of the applications. Computer clocks suffer mainly from **clock skew**, which causes two clocks to progress at different rates, and from **clock jitter**, which causes a clock to tick irregularly.

A good time synchronization protocol has to be efficient and simple but also has to guarantee that:

⇒ Regular synchronization is provided because clocks continue to drift apart after synchronization.

⇒ Available time does not jump back and forth in time because this could result in skipped or duplicated events on the node.

You have also learned about three protocols for time synchronization in sensor networks:

⇒ **Lightweight Tree Synchronization Protocol (LTS)**, which uses the sender-receiver synchronization message exchange, where a sender provides the receiver with the real time.

⇒ **Broadcast Reference Protocol (RBS)**, which uses the receiver-receiver synchronization message exchange and has greater precision than LTS.

⇒ **NoTime**, which does not synchronize the nodes, but instead computes the in-network delay for each delivered data packet, which enables the sink to compute the original event's real time.

QUESTIONS AND EXERCISES

7.1. Recall the broadcast reference time synchronization protocol (RBS) from Section 7.3.2, which discussed that a large number of broadcasts increases its precision. Why does this increase occur? How can this be mathematically shown?

7.2. Discuss how clock skew and jitter affect the precision of the NoTime protocol from Section 7.3.3.

7.3. Recall the sample topology from Figure 7.5. Analyze the work of LTS and RBS based on this topology and compare their performance in terms of the time precision achieved.

7.4. Draw a diagram for NoTime, similar to the ones in Figures 7.3 and 7.4. What is the critical path of NoTime? Is it longer or shorter than for RBT or LTS?

7.5. Recall the hikers from Figure 7.2. Assume that the hikers walk at a maximum speed of 10 km/h and that the sensors are exactly 5 meters apart. What is the required time precision so that events are ordered correctly?

FURTHER READING

Three different time synchronization protocols were presented in this chapter. The Lightweight Tree Synchronisation Protocol is described in great detail in the work of van Greunen and Rabaey [3], while the Broadcast Reference Protocol can be found in the work of Elson *et al.* [1]. There is no dedicated publication for NoTime or a similar protocol, because it is considered too trivial. Some more time synchronization protocols are presented in the survey of Sivrikaya and Yener [2].

[1] J. Elson, L. Girod, and D. Estrin. Fine-grained Network Time Synchronization Using Reference Broadcasts. In *Proceedings of the 5th Symposium on Operating Systems Design and implementation*, OSDI '02, pages 147–163, New York, NY, USA, 2002. ACM.

[2] F. Sivrikaya and B. Yener. Time synchronization in sensor networks: a survey. *IEEE Network*, **18**(4):45–50, July 2004. ISSN 0890-8044.

[3] J. van Greunen and J. Rabaey. Lightweight Time Synchronization for Sensor Networks. In *Proceedings of the 2Nd ACM International Conference on Wireless Sensor Networks and Applications*, WSNA '03, pages 11–19. ACM, 2003.

8

LOCALIZATION TECHNIQUES

Similar to time synchronization, which provides basic time information on WSN nodes, localization provides another important piece of information—the current location of the node. These two services are essential for any WSN application, as they build the basis for the application context, i.e., what happens when and where. This chapter concentrates on localization and explores what types of location information exist for WSN and how to obtain them with some basic techniques.

If you are already proficient in localization techniques for other applications, do not skip this chapter, as localization in WSN can be very different.

8.1 LOCALIZATION CHALLENGES AND PROPERTIES

Location is for humans something as intuitive and natural as time. Just recall one of the typical human nightmares of not knowing where you are and what time we have. For devices, knowing the location is not that intuitive and easily achievable. Most of devices do not know where they are, ranging from washing machines and coffee makers to cars and laptops. Only when a GPS receiver is installed, does a device have location information to use. The following sections discuss the properties and peculiarities of location information for sensor nodes.

Introduction To Wireless Sensor Networks, First Edition. Anna Förster.
© 2016 The Institute of Electrical and Electronics Engineers, Inc. Published 2016 by John Wiley & Sons, Inc.

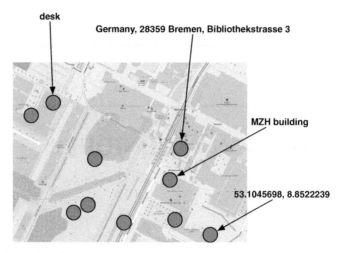

FIGURE 8.1 Different types of location information for outdoor and indoor environments of the University of Bremen's campus in Germany. Some are too coarse while others leave out global information. Adapted from: University of Bremen.

8.1.1 Types of Location Information

First, let us explore what kind of location information is needed. Global addresses, such as postal addresses, are typical for outdoor environments. When these are not available (outside of the postal system), then GPS coordinates are required.

Such location information is also very useful for outdoor sensor networks. But what happens if the complete sensor network is in the same building or building complex? GPS is not available indoors and the postal address is too coarse because it will be the same for all nodes. In this case, you must use semantic information such as the floor or room number of the sensor node. Figure 8.1 shows the difference between possible location types. It shows an excerpt of the campus map at the University of Bremen with some buildings. Some of the shown locations are good enough for outdoor sensors, such as a GPS location in the parking area. However, others may be too coarse, such as the location of the sensor in the MZH building. Where should we look for the sensor? Perhaps in the entrance or somewhere completely else? Other locations may be too precise for some applications, such as the location "desk" for the sensor in building ZHG. There also might be sensors located on different professors' desks but how can you differentiate between them?

symbolic and   Furthermore, you need to note the difference between *symbolic and physical loca-*
physical   *tion*. Physical coordinates remain legal at all times, e.g., GPS coordinates. Symbolic
location   coordinates, such as "desk" can easily change over time, either because the desk is moved away together with the sensor or the desk is moved away, but the sensor node remains where the desk was and gets a new symbolic position, e.g., "chair". Postal addresses are also considered symbolic, even if street names, for example, do not change frequently. However, the largest problem with postal addresses is their

language dependency. For example, in English the country is called Germany, but in German it is called Deutschland.

This problem is also defined as the *scope* of the *location*. If the scope is a single room, then "desk" is perfectly acceptable. If the scope is the campus, then the room number plus the prefix of the building (e.g., room MZH 1100) works fine. If the scope is the world, you would either need to add the address of the building or the GPS coordinates. It is important to decide what the scope is and the kind of location information you need before you start implementing the localization method. **location scope**

8.1.2   Precision Against Accuracy

Different from location scope, localization precision and accuracy refer to how well you are able to localize the sensor nodes. The definition of localization accuracy is as follows:

> **Definition 8.1.** *Localization accuracy is the largest distance between the estimated and the real position of the sensor node.*

As an example, if you use room numbers as location information, you can have accuracy of one room, several rooms, or even several floors. Or, if you use GPS coordinates from GPS receivers, the accuracy is typically 15 meters. This means that any location estimation is expected to be in a radius of 15 meters around the real position.

In contrast, localization precision is defined as follows.

> **Definition 8.2.** *Localization precision is how often a given accuracy is really achieved.*

Continuing with the GPS example, typical GPS receivers achieve 15 meters accuracy with 95% precision. This means that in 95% of all measurements you stay in the radius of 15 meters, whereas in the remaining 5% you have larger errors. This is also depicted in Figure 8.2. From 20 measurements, 19 were inside the expected accuracy radius of 15 meters, while one was outside, which results in 95% precision.

Precision and accuracy are very important metrics for any localization scheme you might use. Even for the simplest one, you need to know how well it does its job.

8.1.3   Costs

Different localization methods have various costs. Some have purely *financial costs*, such as installing a GPS receiver at each individual sensor node. Others have *space costs*, for example, when the GPS receiver cannot fit into the planned space for the

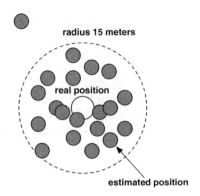

FIGURE 8.2   From 20 localization measurements, 19 were inside a radius of 15 meters around the true position, while 1 was outside. This results in an accuracy of 15 meters with a precision of 95%.

sensor node. These examples are often found in industrial settings, in which nodes have to fit in very small elements such as tubes or machines.

Other possible costs include *communication* or *energy costs*, where the method requires so much communication between the sensor nodes that it becomes unbearable, especially when nodes move around. *Infrastructural* costs refer to the special installation of some infrastructure to localize the nodes, such as GPS anchors.

8.2   PRE-DEPLOYMENT SCHEMES

Pre-deployment schemes include all possible methods to provide sensor nodes with their location information before installing them in the environment. This is very often done for symbolic information, such as furniture items, room labels, or postal addresses. The process is simple yet very error-prone. Every node gets its location manually and you cannot change it later. However, very good deployment planning is needed, where exactly the planned nodes are positioned exactly at the planned positions. Obviously, this method does not scale very well, as it becomes very time consuming to program the locations of hundreds of nodes.

Installing individual GPS receivers on each sensor node is also a kind of pre-deployment solution. In this case, the accuracy is very good (at least for outdoor and relatively large deployments) but the financial cost for acquiring the devices is high. Furthermore, the energy cost for running these devices is also high and often not affordable.

8.3   PROXIMITY SCHEMES

Let us now turn to post-deployment localisation. A simple technique is to guess the approximate location of a node, provided its neighbourhood. Figure 8.3 helps clarify

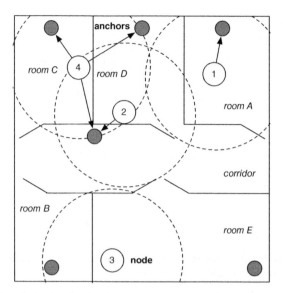

FIGURE 8.3 Proximity localization for sensor nodes. Localization anchors are installed in each room. Depending on which one the sensor node "sees," it simply takes the anchor node symbol or location as its own. However, various problems can occur.

this approach. There is an installed *anchor* in each room, which provides everybody   **localization**
who can hear it with its location. Thus, a sensor node can explore its neighborhood   **anchor**
for anchors and if it sees one, it can assume that it is close enough to have the same
approximate location. For example, node 1 sees the anchor of room A and can assume
that it is also in room A.

The method is not a very accurate. It can also incur various problems and conflict
situations. For example, node 2 sees only the anchor of the corridor and assumes it is
there instead of the correct location in room D. Node 4 sees three different anchors
and will have trouble deciding for one of them. Node 3 does not see any anchor and
cannot position itself at all.

Various remedies have been implemented to tackle the previous problems. How-
ever, *fingerprinting* is considered to be the best one to increase accuracy. After   **fingerprinting**
installing all anchors, measurements are taken at all possible locations (or, at least,
at many locations) and their "fingerprints" are taken. A fingerprint is a tuple, con-
taining all anchors which can be seen at this place and their signal strengths (RSSI or
LQI). Thus, a map of the environment emerges, which can be used by sensor nodes to
compare against and to localize themselves. This approach has two important advantages: it reveals problematic areas of the environment, so that new anchors can be
installed and it enables low-cost localization also for mobile scenarios. However, it
does have high infrastructural costs for the anchors and very high installation costs
for generating the fingerprint map.

Location mules are another interesting approach to proximity schemes. Using   **Location**
location mules, sensor nodes again look for proximity anchors to overtake their   **mules**

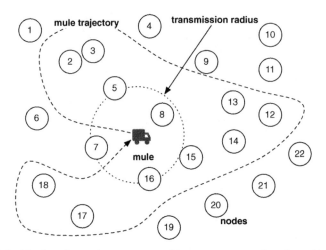

FIGURE 8.4   The location mule moves around in a sensor network and provides the sensor nodes with proximity-based position information.

locations directly. However, there are no installed anchors in the environment. Instead, a mobile node with an on-board GPS receiver is used.

Either a robot or human drives or walks through the sensor network. Sensor nodes can discover the mule when it passes close-by and then take over its current position. The weaker the transmission power of the mule, the better the accuracy of positioning. This approach can also be defined as a pre-deployment scheme because it can also be run before the sensor network starts functioning but after final deployment. It can be run also at any other time, if needed. Figure 8.4 depicts the mule moving around in a sensor network and how it broadcasts its current location to the sensor nodes. The sensor nodes can additionally use the signal strength as an indicator for more accurate position information.

8.4   RANGING SCHEMES

Ranging essentially means measuring the location. For sensor nodes localization, you first do some measurements and then decide on your location. Again, we use anchors and their locations. However, instead of adopting their location as in proximity schemes, in ranging schemes we try to compute our own position depending on the measurements that we do.

Triangulation and *trilateration* are the two main approaches (Figure 8.5). In triangulation, you measure the angle between the sensor node and the anchors and use this information to compute your own position. In trilateration, you measure the distances to the anchor nodes to compute your own position.

The mathematical details for both computations are quite complex so they are not covered here but the interested reader can read Karl and Willig's book [1]. Here, discussion focuses on both approaches' important properties and consequences.

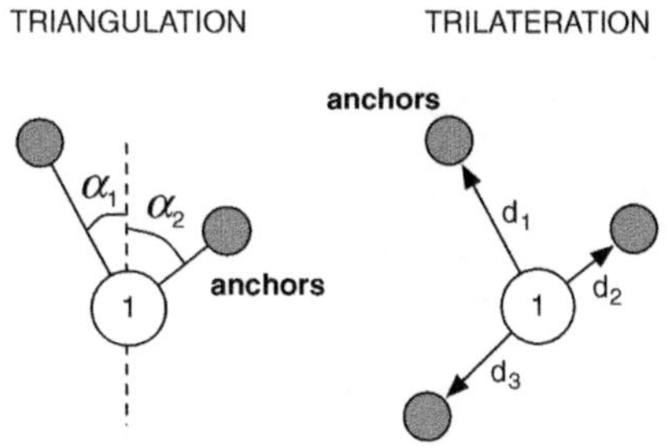

TRIANGULATION   TRILATERATION

FIGURE 8.5 The two main ranging schemes. For triangulation, we measure the angles to at least two anchors. For trilateration, we measure distances to at least 3 anchors.

8.4.1 Triangulation

First, you need to answer the question of how to obtain angle measurements. This is not that simple with standard sensor node hardware, which typically uses omnidirectional radio transceivers. Special hardware is required, such as an array of antennas or microphones on different sides of the sensor node to understand from which *direction* the signal arrives. This approach is costly in terms of hardware but can achieve quite good accuracy.

In terms of the calculation needed, it is relatively simple and does not require any special or costly mathematical functions. It needs a minimum of two anchors for a two-dimensional space. The computation itself is left for homework.

8.4.2 Trilateration

You have already read about some attempts to measure distance based on radio communication such as mapping RSSI values into distances and know this approach is not very reliable. However, others do exist. For example, one could use two different communication interfaces on-board such as radio and acoustic. If the anchor sends a pulse simultaneously through each of the interfaces, neighboring sensor nodes receive them slightly separated in time. Given the propagation speed of both pulses (speed of light for the radio and speed of sound for the acoustic interface), the sensor node can accurately calculate the distance to the anchor. This method is called *time difference of arrival* (Figure 8.6). Keep in mind, the larger the time difference of arrival, the greater the distance to the anchor.

time difference of arrival

Again, this may require additional hardware, which is costly, but not as costly as an array of antennas. In terms of computation, you need more resources to compute the location of the sensor nodes and a minimum of three anchors.

8.5 RANGE-BASED LOCALIZATION

How does a localization protocol work? In general, any localization protocol based on ranging has sensor nodes that look for anchor points and start the ranging

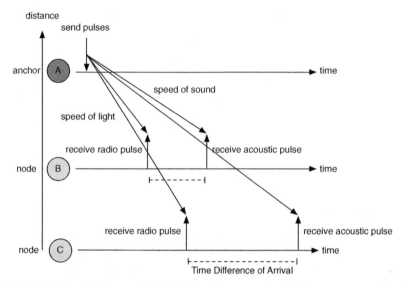

FIGURE 8.6   Time difference of arrival method for measuring the distance to an anchor. The larger is the time difference, the greater is the distance to the anchor.

procedure. If they gather enough anchor points and measurements, they compute their own locations. Once they have localized, they also declare themselves anchors. In this way, sensor nodes without enough anchors get a chance to localize.

iterative   This approach is also called *iterative localization*. The accuracy of localization
localization decreases significantly with each iteration when new anchors are added because the error sum ups over iterations. However, in this way, location information is available at a low price, high scalability, low communication and processing costs.

8.6   RANGE-FREE LOCALIZATION

Range-free localization is a combination of proximity-based and ranging techniques. Strictly speaking, instead of measuring exact distances or angles, this method means trying to guess approximate values and use those for calculating the sensor node's location. The following sections explore some of the most important and widely used strategies.

8.6.1   Hop-Based Localization

In Figure 8.7, three anchors are present in the network and you have their exact positions. Instead of using real distances to the anchors, you can try to approximate how long one hop in this network is. What you already know is the number of hops between the anchors (the bold links in the network) and the real distance between them. For example, the distance between anchors A and B is 130 meters, and they

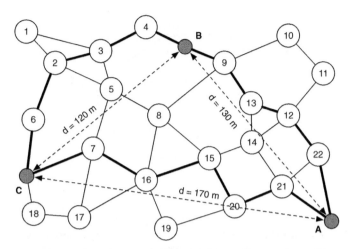

FIGURE 8.7   Hop-based localization of sensor nodes, based on mean length of a hop in a network.

are 5 hops separated. This results in a mean length of 1 hop of 26 meters. The distance between anchors A and C is 170 meters and 6 hops, resulting in a mean hop length of 28.3 meters. The last available distance is between anchors B and C, which is 120 meters and 5 hops, resulting in a mean hop length of 24 meters. You can see that the results are quite close to each other and the network-wide mean hop length can be calculated as $\frac{24+28.3+26}{3} = 26.1$ meters.

The localisation for the sensor nodes works then as follows. It calculates how many hops away it is from each anchor. It uses the mean hop length to calculate the approximate distance to this anchor and can perform trilateration to obtain its own location.

Looking at the topology and real distances in Figure 8.7, it is clear that this approach is not very accurate. The savings from ranging techniques are also not significant, as simple techniques such as RSSI measurements to neighbors are less costly than the mean hop length calculation. However, for really large and dense networks, this approach is well suited.

8.6.2 Point in Triangle (PIT)

Another interesting approach is to compute whether a sensor node is inside a given triangle formed by anchors or not. Figure 8.8 presents the idea of the PIT test (point in triangle). You can test whether a node is inside a triangle by moving it in any direction. If it gets closer/further away from all three triangle corners, then it is outside. Otherwise, it is inside the triangle.

However, it is not practical to move around sensor nodes to better localize them. Instead of moving around, they can use their neighbors as a moved version of themselves. You can simply ask a neighbor for its own measurements to all triangle corners (anchors) and see whether these get closer or further away to all anchors or only to

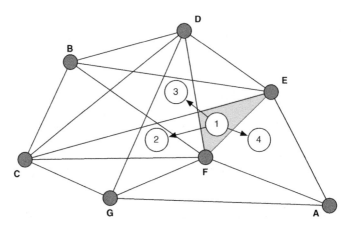

FIGURE 8.8   Point in triangle test (PIT). If the node is moved in any direction and all triangle corners move away/closer, the node is outside. Otherwise, it is inside.

some. Thus, you can identify your own position as inside the triangle or outside. If you are inside the triangle, you can easily calculate its center and assume that this is your own location.

LOCALIZATION: SUMMARY

In this chapter, you learned how to localize sensor nodes. You saw that this problem is quite difficult and cannot be easily solved without investing in infrastructure or additional hardware for the sensor nodes. Location information can be **symbolic** or **physical**, mostly differentiating between human-understandable but not precise information such as postal addresses or more machine-oriented information such as GPS coordinates.

You also learned about several possibilities to tackle this challenge and how each method has its own costs and properties:

⇒ **Pre-deployment methods** include manual or semi-manual procedures of "burning" the right position information onto the right sensor nodes. This is a financially cheap and quite accurate but a very error-prone and tedious process.

⇒ **Proximity-based methods** look for communication neighbors with known positions and simply take this information as their own. This is simple but not accurate.

⇒ **Range-based algorithms** first measure either the angles or the distances to known anchors and then use triangulation or trilateration to compute quite accurately their own locations. The main challenge here comes from performing the measurements because they either require additional hardware (angles and distances) or are quite unreliable (distances).

⇒ **Range-free methods** try to avoid exact distance measurements and approximate them with a mean length of a hop or with simple inside-outside tests. These are not accurate and do not save much energy compared to ranging techniques.

Each localization approach can be evaluated in terms of its **accuracy** and **precision**, e.g., GPS has an accuracy of approximately 15 meters with 95% precision.

QUESTIONS AND EXERCISES

8.1. A sensor node is in reality positioned in coordinates (10, 20). A localization algorithm has calculated in sequence the following locations for it:

X coordinate	Y coordinate
8	19
9	21
10	21
7	18
10	30

What is the precision of the localization algorithm with 80% accuracy?

8.2. Recall trilateration from Section 8.4. Why do you need a minimum of three anchors for trilateration? Draw example topologies and reproduce the computation of trilateration.

8.3. Recall triangulation from Section 8.4. Why do you need a minimum of two anchors for triangulation? Draw example topologies and reproduce the computation of triangulation.

8.4. Recall Figure 8.7 and hop-based localization. Using the facts given in the figure and text, compute the locations of nodes 2 and 10. Now, assume the mean hop length was calculated to be 25 meters instead of 26.1. What are the new locations of nodes 2 and 10? How much did they change from the original calculations?

8.5. Recall the PIT test from Section 8.6.2. Discuss its work in various topologies, e.g., when all neighbors of the node are also inside the triangle, some are inside and some outside, or all are outside. Can you find a scenario in which the PIT test will not work (at least its neighbor-based approximation)?

FURTHER READING

Most of the localization techniques presented here have not been developed exclusively for sensor networks. A comprehensive survey of existing protocols and algorithms can be found in the survey of Mesmoudi *et al.* [2]. The book of Karl and Willig [1] provides a more academic view of the topic and also works out the computation steps of trilateration and triangulation for the interested reader.

[1] H. Karl and A. Willig. *Protocols and Architectures for Wireless Sensor Networks*. John Wiley & Sons, 2005. ISBN 0470095105.

[2] A. Mesmoudi, M. Feham, and N. Labraoui. Wireless sensor networks localization algorithms: a comprehensive survey. *CoRR*, abs/1312.4082, 2013. URL http://arxiv.org/abs/1312.4082.

9

SENSING TECHNIQUES

It is quite peculiar to observe that this book is about sensor networks and that we only start talking about their sensors in this chapter. As you learned in previous chapters, getting the sensed information from the sensor node to the outside world is a very complex and sensitive process so you needed to master those first. Now, you are finally ready to turn to the sensors and data produced by them, their variations and properties.

9.1 TYPES OF SENSORS

There are endless types of sensors available. Some are readily available to be used in particular platforms, whereas others need to be interfaced first to be used in your sensor node platforms. Some are very complex and expensive, whereas others are considered standard and very low price. Some are large and others are tiny. Figure 9.1 offers some examples.

Furthermore, you mostly differentiate between passive and active sensors:

> **Definition 9.1.** *Passive sensors* *passively sense their environment without manipulating it. Very often they also do not need any power to sense their environment – power is only needed to amplify the analog signal.*

Introduction To Wireless Sensor Networks, First Edition. Anna Förster.
© 2016 The Institute of Electrical and Electronics Engineers, Inc. Published 2016 by John Wiley & Sons, Inc.

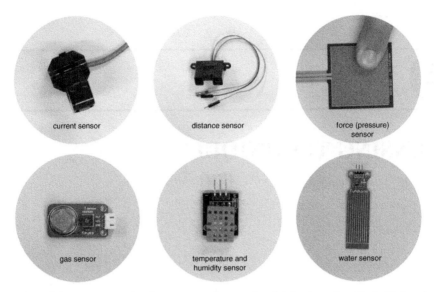

FIGURE 9.1   Examples of various sensors. Source: Special thanks to Jens Dede, University of Bremen.

Definition 9.2. *Active sensors* *need to actively manipulate their environment to sense it, for example, by emitting light or sound waves. Active sensors need power to sense.*

Examples of passive sensors are thermometers, light sensors, and microphones. Active sensors examples are sonar distance sensors or some types of seismic sensors, which generate shock waves.

There is also differentiation between *omnidirectional* and *narrow-beam* or directional sensors. With omnidirectional sensors, the sensed phenomenon does not really have a position or direction, e.g., temperature is sensed omnidirectionally. On the contrary, narrow-band sensors can sense the environment only in one direction or a small angle. For example, sonar distance sensors emit a sound wave in a given direction to calculate the distance to the next obstacle in that direction. An example of a passive, but directional sensor is a camera.

**omnidirec-
tional
narrow-
beam**

9.2   SENSING COVERAGE

How far can a sensor sense its phenomenon? For example, if you have a movement sensor, what is the maximum distance to the movement that the sensor can actually detect? And what happens if this distance is exceeded? These answers are provided by the sensing coverage and *sensing model*. The coverage simply provides you with the maximum distance at which the sensor still works reliably and correctly. The sensing model provides you with information of what happens if this distance is exceeded.

**sensing
model**

Definition 9.3. *The **binary** sensing model assumes that the sensor delivers perfect results inside its coverage area and faulty or missing results outside.*

Definition 9.4. *The **power law** sensing model assumes that the sensor's reliability decreases with increasing distance from the phenomenon.*

Binary and power law are the two main sensing models. A combination of both is of course also possible. In practice, it is preferable to use the binary model based on real observations. For example, experiments can show what the maximum distance is from a moving object so that the movement sensor can detect it. This maximum distance is then considered your coverage area and you rely fully on it.

9.3   HIGH-LEVEL SENSORS

A high-level sensor is a sensor that does not directly correspond to any sensor hardware. For example, if the sensor is movement in a camera image, you cannot simply ask the camera (at least typical cameras) to tell you whether there is a movement in its image or not. Instead, at least several individual images from the camera need to be processed and a decision then made about whether there was some movement or not.

Another term often used for high-level sensors is *sensor fusion.* The term sensor **sensor fusion** fusion indicates that several sensor inputs need to be fused or combined together to receive a new sensor. For example, a high-level sensor could be called "house intrusion" and its low-level sensors could be "door code entered," "movement sensor," "door open sensor," and "window open sensor." All of these four sensors are fused together to obtain the "house intrusion sensor." However, fusion is typically more than just a list of sensors. The decision is usually complex and many different combinations of sensor data and preprocessing are performed. Table 9.1 shows the various possible combinations for the example "house intrusion" for a house with a cat, which can freely move in and out of the house using its cat flap. For example, when the door

TABLE 9.1   Simple sensor fusion with four low-level sensors, combined into one high-level sensor

Sensor	Input						
Door code entered	Yes	No	No	No	No	No	No
Movement	n/a	No	No	No	Yes	Yes	Yes
Door open	n/a	No	Yes	n/a	No	Yes	n/a
Window open	n/a	No	n/a	Yes	No	n/a	Yes
House intrusion	No	No	Yes	Yes	No	Yes	Yes

and the window are closed, but there is movement in the house, the high level sensor still says "no intrusion," since probably the cat is just moving around the house.

Much more complex sensor fusion algorithms are possible and even usual. For example, various data streams often need to be fused such as video with audio. This process is sometimes done on sensor nodes but other times on the server, depending mostly on where the sensors are (all on the same node or not) and how costly the sensor fusion is.

9.4   SPECIAL CASE: THE HUMAN AS A SENSOR

A special type of a sensor is the human itself. The human sensor is the main sensor in crowdsourcing and participatory sensing applications. An example is when a person takes a picture of an event or writes a review of a restaurant. Although these applications are highly fascinating, they are outside the scope of this book.

There is also one sensor on almost any sensor node, which is in fact a human sensor: a button. The button is a very important sensor and can be used in a variety of ways. In some ways, it is also an actuator because what usually follows is an action on a sensor node: reboot the node, send all available information to the sink, call all sensors, etc.

9.5   ACTUATORS

Actuators are similar to sensors but do quite the opposite. Instead of sensing some phenomenon in the environment, they can manipulate it. They can switch on the light or the air conditioning or can start an irrigation system, etc. In terms of usage and programming, they are even simpler than sensors. All you need to do is send them a signal with their new state. Table 9.2 extends the previous example of intrusion detection with actuators. In this example, the following actuators have been added: a loud alarm in the house and a silent alarm, which sends a short message to the homeowner. The loud alarm follows exactly the house intrusion result. The silent alarm is always activated together with the loud alarm but additionally when there is movement in the house.

TABLE 9.2   Simple sensor fusion with three low-level sensors, combined into one high-level sensor

Sensor				Input			
Door code entered	Yes	No	No	No	No	No	No
Movement	n/a	No	No	No	Yes	Yes	Yes
Door open	n/a	No	Yes	n/a	No	Yes	n/a
Window open	n/a	No	n/a	Yes	No	n/a	Yes
House intrusion	No	No	Yes	Yes	No	Yes	Yes
Loud alarm	Off	Off	On	On	Off	On	On
Silent alarm	Off	Off	On	On	On	On	On

Even if the usage of actuators is simpler than the usage of sensors, special attention needs to be given to them. In fact, they manipulate the environment automatically and this manipulation can be dangerous. For example, a small error in the application can cause the preceding house intrusion alarm to sound continuously or an irrigation system to either never irrigate or always be fully turned on. Sensors, on the contrary, do not impact the environment. Data might be lost or erroneous, but this is not dangerous, even if inconvenient.

Another issue with actuators is how to ensure that they actually did their job. Is the window now really closed? Is the heating now really off? Typically sensor data is used to verify the work of the actuators.

9.6   SENSOR CALIBRATION

The values received from a sensor are in fact voltage measurements of an electronic device. Let us consider the example of a thermometer. Perhaps the one you have at home is a small glass one with mercury inside. When the temperature rises, the volume of the mercury increases and its level rises. What you see is actually this level in the context of printed numbers on the glass. Now imagine someone has deleted the printed numbers. How can you know what the temperature is?

Digital thermometers have a similar problem whenever the outside temperature affects the output voltage of the device. All you know is that the voltage is different for different temperatures. But what is exactly the current temperature? The process of finding out is called calibration.

One way to calibrate your temperature sensor is to use an already calibrated one, like your home mercury-based one with a printed scale. You need two measurements, one low and one high. The best choice is the lowest and highest measurements you expect to get. Then, you need to match the received voltages with the temperature you have observed at your home thermometer. Figure 9.2 reflects this. All you need to do now is to read the voltage and transfer it to temperature. Typically, this is done with a table or an equation, not a graph.

Another way to calibrate is to use real physical processes instead of another thermometer. The boiling temperature of water is $100°$ Celsius at 1013 mbar air pressure. The freezing temperature of water at the same air pressure is $0°$ Celsius. Thus, you could wait for a day with approximately 1013 mbar air pressure and prepare one pot of boiling water and one with ice cubes. You can stick your sensor in each of the pots and read the voltage. Then, proceed as in the previous example. The advantage here is that you do not transfer the calibration error of your home thermometer, which has been calibrated with another one and so on. The disadvantage is the error of the experimental setup (or whether your sensor will not be boiled).

Using more than two points to calibrate is advisable in order to verify the obtained graph or equation.

But why do you need to calibrate the sensors when the manufacturer can probably do it better? They actually do calibrate almost all sensors (unless your sensor specification sheet states something else), but the sensor's properties change over

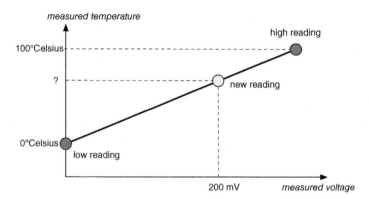

FIGURE 9.2   An example of how to calibrate a digital sensor with a home thermometer. You need two measurements to draw the dependency between the read voltage measurements and the real temperature. Then, any other new reading can be transferred easily from voltage to degrees.

time. The performance of sensors depends on many other components on board the sensor node, such as the battery type used. Even if the manufacturer has calibrated the sensor, you need to repeat this calibration process regularly to ensure correct and reliable sensor readings.

9.7   DETECTING ERRORS

There are two types of error detection: offline and online. Offline detection is relatively simple, even if tedious. The sensor needs to be tested and its output compared to a reliable sensor. This process is very similar to calibration and calibration is also used to detect problems with a given sensor. For example, continuing with the temperature sensor from Figure 9.2, if the dependency line were horizontal, the sensor would have been faulty.

Online error detection is much harder. When the sensor readings change, you have two possible explanations: either the environment has changed or the sensor is faulty. Thus, the main challenge is to differentiate between these two possibilities, which can be difficult.

One option is to always use several sensors, instead of only one and compare their results. If they match, all sensors work correctly. If they do not match, at least one of the sensors is faulty and this can be signaled to the sink. Another option is to observe the current battery voltage on the sensor node. As Section 2.2 discussed, different components on the node require different input voltages. Thus, while other components might be still working correctly, a sensor might not work reliably any more at lower voltages. Finding out the minimum required voltage and observing the currently provided one is an efficient way to recognize potentially faulty readings.

SENSING TECHNIQUES: SUMMARY

Sensors build the most important part of sensor networks. Without sensors on board, a sensor network is only a toy or at best a prototype. Different types of sensors exist such as **active** and **passive** sensors, **omnidirectional** and **narrow** band sensors. All sensors can be characterized by their:

⇒ **Sensing coverage**, which defines the maximum distance and area at which the sensor works reliably.

⇒ **Sensing model**, which defines how reliability changes with increasing distance from the sensor.

In practice, binary sensors are preferred, which work reliably in a predefined coverage area and do not work reliably outside of it.

The process of matching the output of a sensor to some physical entity (such as temperature in degree Celsius) is called calibration. Sensors need to be calibrated periodically to ensure that they still deliver correct results.

Sensor errors can be detected offline (before deployment) in the process of calibration or online (during deployment). The second is challenging due to changing sensor readings, which can come from changes in the environment or faulty sensors.

QUESTIONS AND EXERCISES

9.1. Imagine a traffic light management system as shown in Figure 9.3. There are sensors on each of the streets in both directions (8 sensors total). Each of them has three possible states: no cars, few cars, and many cars. There is also one actuator, which switches the traffic lights. It has two possible states: "left to right" and "up to down." The meaning of these states is that the traffic lights are either green to the traffic going horizontally in the figure or vertically. Assume nobody turns at the crossing, all cars go only straight and need to cross.

Create a table with the possible states of all 8 sensors and decide on the new state of the actuator, given the sensor readings. You can assume that the sensor readings arrive periodically and that the traffic lights switch with a pause in between to allow the cars already in the crossing to cross completely.

9.2. In Section 9.2, we have discussed two different sensing coverage models – binary and power law. However, we also said that in practice we prefer to use the binary model. Discuss this statement and compare the usage of both models in practical applications. Consider the following example: you are sensing the temperature of an area of 10 × 10 meters with a regular grid of sensors every 2 meters. The binary sensing model says that the sensor works reliably with a

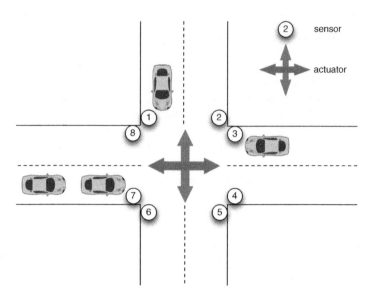

FIGURE 9.3   Traffic light example with 8 sensors for each lane in each of the four streets and one actuator.

coverage radius of 2 meters. The power law sensing model says that the sensor's reliability starts decaying after 1 meter exponentially (assume whatever exponent). How can you use the different models to reconstruct the complete temperature map of the environment? Discuss advantages and disadvantages of both models.

9.3. Sensor readings are highly dependent on the battery's voltage. Prove this experimentally. Take a sensor node and put fresh batteries into it. Put the node into a room with a relatively stable temperature. Then, read the temperature continuously and save the values into the on-board log. Be careful not to connect the node to a USB cable, as it will provide it with external power. Leave the node until it dies and then read and plot the measured values.

9.4. Calibrate the temperature sensor of your sensor nodes as described in Section 9.6. Use a home thermometer to calibrate them. Plot the resulting graph and transfer the dependency into an equation. Now make some further measurements to verify the calibration.

10

DESIGNING AND DEPLOYING WSN APPLICATIONS

So far you have learned various individual challenges arising in sensor networks and how to solve them. This chapter helps you develop a systematic approach toward designing and deploying complete wireless sensor networking applications. It provides you with the basic steps needed and discusses their properties, challenges, and advantages.

Why is such a systematic approach needed and why spend an entire chapter on it? Wireless sensor networks are still in their baby shoes and their development and deployment is a complex and hard task. However, most problems and challenges are caused by the environment where they are deployed and by insufficient preparation of the deployment. It is like building a spaceship without knowing the weather in space.

10.1 EARLY WSN DEPLOYMENTS

Many researchers have already identified and discussed the problem of deployment such as in the surveys of Beutel *et al.* [1] and Oppermann *et al.* [4]. These surveys also hold more information about recent successful deployments. There are two prominent examples for failing deployments: the so-called Potatoes application and the Great Duck Island application. Both applications are from the beginning of the sensor networks era, but they are useful to analyze caveats to unexperienced practitioners.

Introduction To Wireless Sensor Networks, First Edition. Anna Förster.
© 2016 The Institute of Electrical and Electronics Engineers, Inc. Published 2016 by John Wiley & Sons, Inc.

10.1.1   Murphy Loves Potatoes

This application, officially called LOFAR-agro [3], was developed for precision agriculture for potato fields in the Netherlands to monitor temperature and humidity, and to take centralized irrigation decisions based on the acquired data. The resulting publication [3] is worth reading, especially because the authors describe in a honest and direct way their disastrous experience from a project, which at the end managed to get only 2% of the targeted data for a much shorter period of time than targeted.

The problems were numerous. After careful testing of the nodes in the lab, an accidental last-minute update to the code of the MAC protocol caused it to only partially work and not deliver any data at all. Yet another problem was caused by the code distribution system installed, which caused the nodes to continuously distribute new code throughout the network and depleted the nodes' batteries in only 4 days. The next revision brought up a problem with routing, as it was trying to maintain a table of all neighbors, sometimes up to 30. However, such large routing tables were not planned and could not be stored, resulting sometimes in no routing paths available to the sink. An attempt to manage node reboots wisely and the addition of a (faulty) watchdog timer resulted in most of the nodes rebooting every couple of hours. In summary, instead of running for more than 6 months (the complete potato growing period), the network ran for only 3 weeks and delivered 2% of the gathered data during this short period.

Node reboots, lifetime problems, and faulty code were the problems in this example. Later sections discuss each of these elements and how they can be avoided.

10.1.2   Great Duck Island

This is one of the first-ever WSN applications for environmental monitoring. In 2002, 43 nodes were deployed on Great Duck Island, USA, to observe the breeding period of Leach's Storm Petrels [5]. The nodes were equipped with infrared, light, temperature, humidity, and pressure sensors. Once installed, researchers were not allowed to go back to the island because they would disturb the island birds while they were breeding.

Again, the problems were numerous, but this time caused mainly by weather and environmental conditions resulting in hardware faults. A crash of the gateway caused the complete data of two weeks to be lost forever. Water entering the casing of the nodes interestingly did not cause them to fail, but for the sensors to deliver faulty data such as humidity readings of over 150%. These errors triggered an avalanche of problems with the other sensors because they were all read through the same analog-digital converter. However, communication problems were actually rare because of the network being a one-hop.

In 2003, a second attempt was made but routing was introduced and the casings were updated. The trial was also divided into two batches with 49 nodes deployed as a one-hop network, similar to the year before. Batch B included also routing and consisted of 98 nodes. Again, the gateway failed several times causing much of the data to be lost. The node hardware performed better this time, but routing caused overhearing problems at the nodes, which was not previously tested in the lab. This caused many nodes to die prematurely. In the end, the first 2002 deployment delivered approximately 16% of the expected data and the second trial delivered 70% for the batch A network but only 28% for batch B. The ratio of gathered data against expected data is also called the *data yield*.   **data yield**

Note here the different set of problems. Hardware issues and maintenance unavailability were great enemies. Furthermore, the researchers added unnecessary complexity to the network such as the unneeded routing protocol. Semantic problems are one of the most difficult problems for WSN practitioners, as the data gets delivered successfully but it is completely useless. This typically renders all of the data useless, as no way exists to identify which readings are faulty and which are not, so they all need to be considered faulty.

10.2 GENERAL PROBLEMS

Looking at the experiences of the WSN researchers from the previous examples, what is the definition of a problem? A problem is an unexpected behavior of a node or a group of nodes, which does not correspond to your formal or informal specification. The following differentiates problems in more detail, loosely adopting Beutel *et al.*'s [1] classification of problems.

- **Node problems** occur on individual nodes and affect their behavior only.
- **Link problems** affect the communication between neighbor nodes.
- **Path problems** affect the whole path of communication over multiple hops.
- **Global problems** affect the network as a whole.

These problems are caused by varied processes and properties in the environment, but mainly by the following:

- **Interference** in the environment causes various communication problems in the network as a whole.
- **Changing** properties of the environment, such as weather conditions, furniture, or people density, causes various communication problems but also other hardware problems such as complete shutdowns due to low temperature.
- **Battery** issues cause not only problems at the node level but also for communication and sensing as hardware performance deteriorates slowly with decreasing battery levels (see Chapter 2).
- **Hardware reliability** causes various global problems such as communication issues, sensing issues, but also node reboots.
- **Hardware calibration** is a typical problem for sensing because it delivers wrong sensor data.

- **Low visibility** of the nodes' internal states does not directly cause problems but favors unrecognized errors until late in the design process, which prove to be hard to detect and resolve.

The following sections explore the individual problem levels in more detail, before we discuss how to avoid them in the design and deployment process.

10.2.1 Node Problems

Node death *Node death* is one of the most common problems. Suddenly, a node stops working completely. This might seem like a serious problem and it is, but there is also some good news. When a sensor node suddenly stops working, its internal state is still valid. This means that whatever data already exists on this node, it is still there and valid (unless the sensor does not store data at all). Also, neighbor nodes are usually well prepared for dead nodes and quickly exclude them from routing or co-processing. Thus, real node death is a problem, but a manageable one. However, what sometimes looks like a dead node is often not a completely dead one. This can be due to various reasons such as:

- **Slowly deteriorating batteries** causing individual components on the node to perform badly or not at all, while others continue working.
- **Faulty components** causing errors while sensing, storing, or communicating, while the general behavior of the node is not affected. Faulty components could be both hardware and software.
- **Node reboots** typically take time and cannot be easily detected by neighbors. They are often caused by software bugs. During this time, the node seems to be unresponsive and uncooperative.

These problems lead to unpredictable node behavior, which results in semantic errors in the application. This means that the data coming out of the network does not make sense any more. These errors are extremely hard to detect and localize, as they **seemingly** are unpredictable and indeterministic. Thus, the *seemingly dead nodes* are by far the **dead nodes** largest problem in WSN deployments.

> *Fighting seemingly dead nodes.* In order to mitigate these problems, easy steps can be taken to observe the individual components and to report or save their states for later inspection. By far the simplest remedy is to count the number of reboots and to observe the battery level. Both events can be additionally signaled via special messages or special fields in regular messages to the sink of the network.

More sophisticated strategies can also track the state of individual hardware and/or software components such as radio or sensors. They are often programmed to report problems by returning some sort of an error (e.g., `false` to a send packet request of the radio). However, these more sophisticated techniques require more memory and processing and they can also overwhelm the primary application.

Other node problems include various software bugs. These lead to clock skews, which lead to unsynchronized nodes and missed communication opportunities, hanging or killed threads, overflows in counters, and many other issues. It is essential to test the software incrementally and thoroughly to avoid the manifestation of these problems at the deployment site.

10.2.2   Link/Path Problems

Obviously, node problems also cause link and path problems. General propagation or path loss is also a problem because nodes are prohibited to communicate where expected. However, the biggest issue is *interference*. Interference can occur between **interference** individual nodes in your network but also between your nodes and external devices such as mobile phones, Bluetooth devices, or a high-voltage power grid. You cannot see the interference and it changes quickly and unpredictably.

In fact, interference by itself causes "only" lost packets. However, these sporadically lost packets trigger an avalanche of other problems at all levels of the communication stack. They cause more traffic because packets need to be resent. They also cause link prediction fluctuations because the link protocols believe something happened in the network. They force the routing protocol to change its paths and cause the application to buffer too many packets, leading to dropped packets. At the routing level, these fluctuations might lead to temporarily unreachable nodes or loops in the routing paths. These, in turn, cause even more traffic and can make the network collapse completely.

> *Fighting link and path problems.* Interference cannot be fully avoided, but it can be carefully studied and proper channels selected. However, the far most important task is to ensure that communication is self-sustainable, self-adapting and robust.

How can you achieve robust communications? The employed protocols must be simple and leverage all the communication opportunities that they have. For example, a highly agile link quality protocol, which supports even high mobile nodes, may be a very interesting academic topic but will not work in reality. It will be too fragile for highly fluctuating links under heavy interference. At the same time, another link predictor, which is rather slow in recognizing new or dead nodes, will perform more robustly.

At the same time, we should not forget why we always target better protocols and algorithms. They are able to save a lot of energy and significantly increase the network lifetime. As always, you need a tradeoff between robustness and energy saving.

Other link problems also exist, including traffic bursts, which cause links to congest and packets to get dropped. Asymmetric links lead to fake neighbors. Both can be mitigated in the same way as general link and path problems by designing robust communication protocols.

10.2.3 Global Problems

These problems affect the network's general work and can be coarsely classified into topology, lifetime, and semantic problems.

Topology problems refer to various peculiarities in the communication topology of the network, which lead to errors or other unexpected behavior. *Missing* **Missing** *short links* are links between very close neighbors that are not able to commu-**short links** nicate with each other. These links are expected and planned by the designer to **unexpected** offer alternative communication paths. The opposite problem, *unexpected long links*, **long links** occur when two neighbors with an unexpected long distance between them are able to communicate well. Both problems cause premature battery depletion of some nodes.

Another difficult topology problem caused by missing links is *partitioned net-* **partitioned** *works*. In these networks, nodes in one partition are completely isolated from the **networks** nodes in other partitions. This causes the partition to spend a lot of energy in search of a path to the sink and never send its data.

These problems are depicted in Figure 10.1. The missing link between nodes 1 and 2 causes all the traffic of node 2 to go through node 3, instead of going directly to the sink. Thus, node 3 experiences congestion and will use its battery much faster than desired. The unexpected long link between nodes 11 and 3 causes even more traffic to pass through node 3. All of the traffic from node 9 and 11 will use node 3 instead of nodes 8 and 7. Finally, the partition of nodes 4, 5, and 10 cause these nodes to never deliver any data and waste their batteries searching for routes.

> *Fighting topology problems*. The final topology at the deployment site needs to be stored and analyzed. A simple experiment is sufficient to gather the link qualities of all links in the network.

Lifetime problems refer to an unexpectedly short network lifetime. There are sev- **network** eral definitions for *network lifetime*, but the one that makes most sense in the context **lifetime** of real applications is the time when useful data is gathered. This time might start

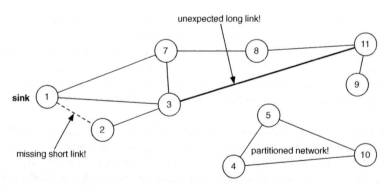

FIGURE 10.1   Topology-related global problems: missing short links, unexpected long links, and partitioning of the network.

immediately after deployment or some time after it. It ends when the network is not able to retrieve as much data as it is needs to semantically process it. Obviously, this definition will have different interpretations in different applications. For example, if you have a plant watering application, in which the humidity level of various pots is monitored and reported, the information remains useful even if some of the nodes do not work anymore. However, if you have structural monitoring of a bridge, you need all sensors to be working properly and reporting data, otherwise the structural integrity of the bridge cannot be guaranteed.

Lifetime problems occur most often when individual nodes use their energy too quickly and deteriorate the network's general performance. The most endangered nodes are the ones close to the sink or at communication knots (they serve as forwarders for many other nodes). However, software bugs also can cause them to stay awake for extended periods of time and waste their energy. Lifetime problems are hard to predict or test because they manifest themselves too late in the process.

Fighting lifetime problems. The most efficient way to predict lifetime problems is to log the awake times and the communication activities of individual nodes over some extended period of time (e.g. a day). This information can then be used to analyse the behaviour of all nodes and to find potentially endangered ones.

Semantic problems occur most often when the data and its meaning does not match expectations. This can be caused by missing or bad calibration of the sensing hardware; by deteriorating batteries causing wrong sensor readings; by wrong interpretation of the data itself; or, by overflowing counters and other pre-processing measures on the nodes. While these problems are probably the most devastating ones in the context of the application needs, they are also relatively easy to test.

Semantic problems

Fighting semantic problems. After final deployment, it is essential to present a data excerpt to the final user of the application and to ask him or her to analyze it thoroughly.

10.3   GENERAL TESTING AND VALIDATION

Whatever the purpose of an application or the target of a study, there are several testing methodologies applicable to WSNs. The main principle of testing and validation is to:

Implement the application in several steps, incrementally increasing the complexity of the testing environment.

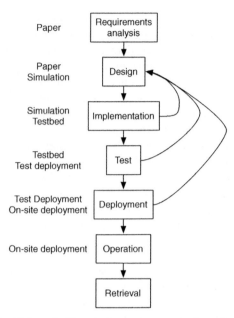

FIGURE 10.2   Typical life cycle of a wireless sensor network application. For each phase, the testing and validation methods are also provided. For example, while the requirements analysis is done mostly on paper, the implementation is tested in simulation and on testbeds.

Figure 10.2 depicts the typical life cycle of an application. The implementation process itself is divided into the following five steps:

1. Start on paper with theoretical models to validate the general idea and algorithms.
2. Proceed with the implementation and simulate the system.
3. Increase the complexity by moving from simulation to a testbed deployment.
4. Further stress the system by introducing all real-world properties in a test deployment.
5. Finally, you should be ready to face the normal operation of the system in its final deployment site.

The previous steps can be taken one-by-one or by skipping some of them. However, skipping any of them either makes the next step more complex or does not sufficiently stress-test the system and thus does not prove that it is ready for operation.

Next we will discuss of the previous life cycle steps in detail. To render the discussion more concrete and realistic, a simple application will be used as a sample scenario called vineyard monitoring. The targeted deployment is depicted in Figure 10.3. The sink is a bit further away in a small house. Each of the sensors in the vineyard

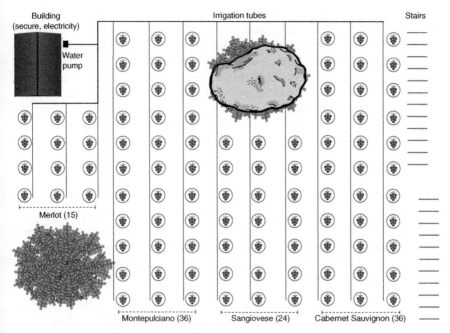

FIGURE 10.3   A vineyard with a WSN deployment.

is expected to gather data for humidity, temperature, and solar radiation every 5 minutes. You can assume that these are the requirements from the vineyard owner. Your task is to design and deploy this network as systematically as possible.

10.4   REQUIREMENTS ANALYSIS

The main goal of the requirements analysis is to ensure that you understand the application, the environment in which it will be running, and expectations of the user. In this case, the user is the owner of the vineyard who is probably not a technician. However, he or she is an expert in the field of growing and producing wine and expects the new system to simplify and support his or her work. In order to find out how your future application will implement the envisioned application, you need to ask several questions and carefully analyze the environment. The questions you need to ask can be divided into four categories: environment, lifetime and energy, data, and user expectations.

10.4.1   Analyzing the Environment

This is probably the most challenging and blurry task. You need to analyze the weather and its fluctuations, surrounding flora and fauna, expected interference, and topology of the site itself.

Weather. The vineyard owner would like to use the sensor network throughout the whole year, even in winter. The owner suspects that the temperature and humidity during winter impacts the quality of the wine in the next season. The vineyard itself is situated in southern Switzerland, close to the city of Mendrisio. Thus, you can expect temperatures from –20 degrees Celsius in the winter (happens rarely, but happens) up to +50 degrees in the summer time. In terms of humidity, all levels from 0 to 100% can be expected. Furthermore, snow and rain are typical for winter and spring and may last for several days in a row. Full snow coverage is possible for extended periods of time (weeks). This data is best obtained from meteorological stations near the deployment site. *Consequence: Packaging needs to be very robust and the hardware itself needs to be tested in extreme conditions.*

Flora. During the normal yearly lifecycle, the flora at the deployment site will appear and disappear. In winter, almost nothing will be left while in summer time the vines will be growing. However, you can expect that no other abundant vegetation will be present, such as big bushes or big trees, as those impact the growth of the vines. *Consequence: Nodes should not be buried under the leaves and should have a similar micro-climate around them.*

Fauna. The vineyard is open and not protected from wild animals or dogs. Domestic animals such as cows or sheep are not expected but also not completely improbable. *Consequence: Packaging needs to be robust against wild animals.*

Interference. There are no high-voltage power transmission grids around the vineyard. In terms of other wireless devices, the only wireless network is the one installed by the owner for home use. However, hiking paths are situated close-by and interference from various Bluetooth or wireless devices can grow during high hiking season. *Consequence: Communications should be carefully tested under high interference and channels with low interference should be selected.*

Topology restrictions. The vineyard is situated on a slope but in general accessible from all sides. However, a big rock is situated in the middle of it, where several vines are missing (Figure 10.3). *Consequence: No sensor nodes can be placed on the rock directly, thus communication around the rock need to be carefully tested.*

Security concerns. These can be both in terms of physical security and cyber security. Tourists typically do not enter the vineyard but it still happens, especially young people at night. The hiking paths are also close so cyber attacks are not impossible. *Consequence: Robust installation, which does not allow easy removal or displacement of nodes.*

Existing systems. Very often, existing systems can be leveraged for a new sensor network. For example, in the vineyard you see a sophisticated irrigation system, which spans the complete area.

In summary, the vineyard application is not extremely challenging, especially because it is easily accessible and weather conditions are not extreme. More extreme environments, such as glaciers or volcanoes, require special measures.

10.4.2   Analyzing Lifetime and Energy Requirements

Next, you need to understand for how long this installation is supposed to work.

Expected lifetime. The vineyard owner needs at least two years of data, the longer the better.

Maintenance opportunities. The owner is regularly checking his vines approximately once a week, which will give you the opportunity to exchange batteries at least once every one or two months. Additionally, since the vineyard is publicly accessible, you can easily enter and test the system without the help of the owner (rather a perfect case and rarely seen).

Energy sources. Solar batteries are a good option, as sun is abundant on the vineyard, even in winter periods. Given the extended total lifetime, they are probably the better and cheaper option.

10.4.3   Analyzing Required Data

Sensory data includes temperature, humidity, atmospheric pressure, wind speed, precipitation, solar radiation, and ultraviolet radiation. These are typical for agricultural monitoring and are all rather low-data rate sensors.

Tolerated delay. The tolerated delay for the data is quite high in winter when no vines are growing and data gathering is only happening for statistical purposes. In this period, even days and weeks are acceptable. However, during vine growing periods, the delay is approximately 15 to 20 minutes so that the sensor network can be directly attached to the irrigation control.

Precision and accuracy. The precision of the data is expected to be normal, e.g., 1 degree Celsius for temperature, 1% for humidity, etc. The data's accuracy should be quite high, meaning that sensor readings should be collaboratively preprocessed to assure that no faulty readings are sent to the user.

Time precision. Time is a critical issue for WSNs because the hardware typically does not have real-time clocks. What is natural to have on a personal computer becomes a real problem to obtain on a sensor node. Thus, you need to understand what the application's requirements are and how much attention you should give to time synchronization. In the vineyard, as in almost any other WSN application, time is important. The time precision, however, is rather coarse and a precision of several seconds to even a minute is acceptable.

Locality of data. This requirement refers to the question of whether you need to know exactly where the data comes from or whether this is not important. For the vineyard, it is important to know whether the data comes from node 1 or 51, as the irrigation system will start watering the area around the node. For other applications, e.g., proximity recognition between two people, it is not important to know where the people are, only that they are close to each other. Such applications are called *location-agnostic*. | location-agnostic applications

10.4.4   Analyzing User Expectations

The vineyard owner has great plans for this installation and would like to better understand the topology of the wine fields and which sub-fields deliver best results. The owner would like to lower the use of water and pesticides, and even attempt to transfer to organic production. Furthermore, the owner would like to include later other sensors such as chemical sensors or biological sensors to early identify problems and sicknesses. Currently, testing the system in one of his fields is of interest but optimally the owner would like to extend the installation to all fields and have all data in one place. Thus, scalability and extendability are important requirements that are best taken into account early.

10.5   THE TOP-DOWN DESIGN PROCESS

The design process targets a complete architecture of the sensor network, including where the nodes are, which protocols to use, and many other factors. However, it does not go into implementation details. Design and implementation are sometimes hard to differentiate. For example, in the design process the designer could decide to use a particular routing protocol and MAC protocol. However, both protocols are complex and both cannot fit on the sensor node. Is this a design problem or an implementation problem? In fact, it is considered a design problem. The design should make sure that all these details are discussed and evaluated, and that the complete architecture of the system is well analyzed by system experts.

Section 10.6 focuses on the implementation process and deals with connecting individual components and mainly testing them.

10.5.1   The Network

Given the vineyard and the requirements analysis, you have several options for the general network. You could install a sensor at each vine, having a total of 111 nodes. This is quite a lot for such a small vineyard. At the same time, this deployment will also provide the highest sensing precision.

Another possibility is to deploy the network only in half or one-third of the vineyard, e.g., in the half closer to the house. This will provide the owner with high precision and a good point for comparison with the other half. Yet another possibility is to deploy sensors not at all vines, but only at some of them. Here, the experience of the owner is crucial because the sensors need to be placed at the most relevant positions. In this case, these are the places where the land flattens (in the middle of the vineyard, where the stairs are interrupted), and around the trees, as there the land is able to store more water and the temperature is lower. Furthermore, you need to consider that different grape vine varieties (Merlot, Sangiovese, etc.) have different properties and different needs. After a discussion with the owner, it makes most sense to install sensors at approximately every second vine in each direction, as shown in Figure 10.4.

The network consists of 36 sensor nodes, each of them gathering temperature, humidity, and light readings. The sensors are positioned to cover the flat areas and different grape vine varieties. Furthermore, the neighboring sensors between

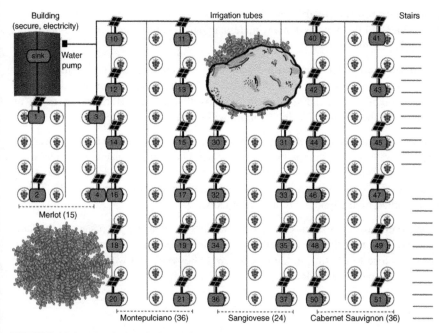

FIGURE 10.4   The designed vineyard sensor network at network and neighborhood levels.

Montepulciano and Sangiovese, for example, offer opportunities to see fine-grained differences between the grape vine varieties and between neighboring vines.

Instead of considering exchangeable batteries, you should consider solar powered ones as previously discussed. Each sensor will be equipped with a small solar panel, similar to those used for solar garden lamps. The sink is positioned in the house, which is connected to the Internet and power grid.

The sink does not need any special protection because it is in a well-protected house with stable temperature and humidity levels. However, the sensors need protection from animals and curious tourists. They will be installed on pillars as high as the vines and firmly connected to them. They are also protected with transparent, but hard packaging and have a QR code to explore. This typically works well, as most of the curious people just want to know what this thing is. A QR code offers a possibility to explore without damaging the sensor node, intentionally or not.

10.5.2   The Node Neighborhood

In terms of the network topology and the neighborhood properties, since the network is relatively dense and distances short, problems should not be expected. In order to fully secure this, you should perform a couple of experiments in the field and test both short and long links. The problems you are typically looking for are:

Isolated nodes.   In this case, nodes 41 and 51 might be isolated, as they are the outermost nodes. It needs to be tested whether they are connected to at least two neighbors.

Nodes with too few neighbors. This is similar to the previously isolated nodes but less critical. In this vineyard, many nodes can have in theory only one neighbor, e.g., nodes 41, 20, and 51.

Chains of nodes. The previous problem often leads to chains of nodes. In this vineyard, the probability of having chains is low because you have installed a grid of nodes. In other cases, chains are very risky, as their entire performance depends on all nodes in the chain and a single failure will break the complete chain.

Nodes with too many neighbors. This problem is quite opposite from the previous ones. However, when a node has too many neighbors, it also needs more energy to maintain information about them and it might serve as a forwarder for too many nodes. In this case, many nodes could have many neighbors, e.g., node 32. This is best handled by the link and routing protocols.

Critical nodes. These nodes serve as forwarders for many neighbors and at the same time have few neighbors to reach the sink. In this case, many nodes (1, 3, 12, and 10) are very close to the sink so the risk is low. Otherwise, you would need to provide extra nodes to spread the load.

10.5.3 The Node

As we discussed in Section 10.2.1, the main node-level problems are dead nodes and seemingly dead nodes. What is important at this stage of the design process is, what will happen if some of the nodes die? If the previous neighborhood-level design was properly done, nothing bad should happen if some of the nodes die. However, you can still differentiate between two types of networks:

- One-hop networks. In these networks, the death of individual nodes does not affect the work of the other nodes. The network can die gradually and even if only one node is left, it can work correctly. Thus, node dying is "allowed" and does not need any special precautions in the implementation phase.
- Multi-hop networks. In these networks, the death of individual nodes affect the work of others and especially their communication. Even if some in-network cooperation is performed before sending the final result to the sink (e.g. two nodes agree on a temperature value), the network can still be considered a multi-hop. Here, you cannot really allow for too many nodes to die and you need to take implementation steps to recognize and prevent the death of nodes.

In this vineyard, we have a second type of network and thus we need to recognize and prevent node deaths.

10.5.4 Individual Components of the Node

From the previous discussion, it is time now to derive which components are needed in the vineyard. Figure 10.5 presents a summary of the needed components. It may

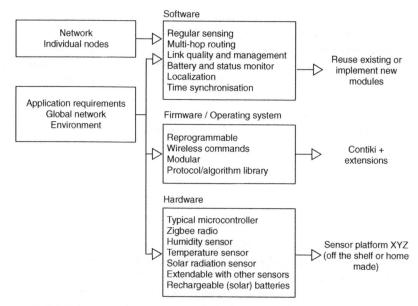

FIGURE 10.5   The derived required components for the vineyard application.

be quite disappointing to see that all you really need is Contiki, with some extensions to it and six rather standard protocols/modules. However, this is the greatest advantage of a structured analysis and design. You are left with only the most necessary modules, without including anything you do not necessarily need. The next section discusses some of the most important concepts and approaches when implementing the vineyard application.

10.6   BOTTOM-UP IMPLEMENTATION PROCESS

The implementation process starts bottom-up and follows Figure 10.5. However, first you need to identify the hardware to use. Essentially, you have two options: off-the-shelf or custom-built (home-made) nodes and solutions.

For straightforward and rather standard applications we will be often able to find a complete solution on the market, e.g. sensor nodes already connected to the cloud. These look tempting and in some cases they might be the perfect solution. However, the following properties need to be evaluated for both off-the-shelf and custom-made solutions:

- **Price.** You can expect the price of a complete solution to be much higher than for home-made solutions.
- **Requirements fulfillment.** Are all your requirements really fulfilled? You need to check carefully all items in Figure 10.5. For example, some off-the-shelf

solutions might not offer solar batteries. Furthermore, the use of data and their representation might be different from what the user expects.

- **Programmability.** Can you change the software or is it proprietary? This is very important when the application or some of the software components are rather non standard. This is often the case with consumer-oriented products, such as smart home systems.

- **Extendability.** Especially when the user would like to extend the system later, it is crucial to select a system, which can be freely extended and modified. This is clearly given in home-made systems but often a trouble with off-the-shelf solutions.

- **Service agreement.** Off-the-shelf solutions often offer paid service agreements, which guarantee the work of the system to some extent. This is convenient but also inflexible.

If you decide on a complete off-the-shelf solution, you may close this book now. But the opposite decision, making everything from scratch including the hardware is also not in the scope of this book. The typical decision is to select an off-the-shelf hardware, which can be programmed by open source operating systems and architectures such as Contiki. This should also be your decision for the vineyard application.

10.6.1   Individual Node-Level Modules

You can now proceed with the implementation of the individual components, as shown in Figure 10.5. This book does not discuss the source code and the exact algorithms because they are too application-centered. However, some important concepts need to be considered:

Define first the input and the output of individual modules. For example, for your regular sensing component, the input is a timer firing regularly and the output are these values: temperature, humidity, and solar radiation.

Reuse as much as possible. If existing modules already offer the required functionality, reuse them. If there is already a module for regular temperature sensing, reuse it to add humidity and solar radiation readings.

Ensure you fully understand reused modules. For example, when you decide to reuse the temperature regular reader, make sure it does exactly what you expect it to do. It could be that it only delivers the temperature if it is above some threshold.

Do not make dependencies between the components. This is one of the most important concepts as it allows you to test individual components separately from each other and reuse them for other applications. Do not make any assumptions about which module does what, but implement the module in a self-sustainable way. For example, the link quality protocol can reuse packets arriving for and from other modules, but it should also be able to create its own packets, if no others are present.

The testing of the individual node-level modules can be best done in simulation, where many different scenarios and event sequences can be tried out.

10.6.2   The Node As an Entity

Once the individual components are implemented and tested, you should analyze the node as a whole. The main principles are as follows:

- **Make sure you have enough memory/processing.** Individual components rarely have a memory problem. However, putting all components together might result in a memory and/or processing bottleneck.
- **All components are correctly interconnected.** Thoroughly analyze the interconnections between the components of the system. Look especially for shared data structures, where reading and writing accesses might create collisions.
- **Event order.** Sometimes the events in a system might be ordered in a weird way. Consider the following example: perhaps you use two timers, one for reading the sensory data and the other for sending the packet with the data out. However, sometimes the reading timer fires first but sometimes the sending timer does. If this order is not deterministic for some reason, the packets will sometimes hold duplicate data and sometimes skip readings. The event order is crucial for making the node as a whole work correctly.
- **Reboot handling.** What happens when a node reboots? How does it obtain correct internal and external states? You need to make sure that the node always starts clean and that it is able to obtain a correct state either from internally saved information or from communication with others.

The testing of the node as a whole can be done again in simulation. However, testing on hardware is crucial at this step. For example, a shared testbed such as INDRYIA [2] can be used.

10.6.3   The Network As an Entity

Finally, you need to analyze the behavior of the network as a whole. This relates mostly to communication and topology issues and the main concepts are as follows:

- **Reboot handling.** Again, a node reboot might result in unexpected behavior, wrong routing information, time, and location. What happens when a node reboots? How do the other nodes react to this?
- **Communication stands.** Make sure that the implemented communication protocols work correctly to avoid collisions, congestions, and energy wasting.

Testing of the complete network can be conducted again either in simulation or in testbeds. However, it is crucial to confirm the implementation in a sample deployment site or in the final deployment site. The choice often depends on costs and the availability of testing possibilities. For example, testing a parking system on a university or industrial campus is feasible, whereas testing the vineyard on a campus is not.

DESIGNING AND DEPLOYING WSN APPLICATIONS: SUMMARY

What did you learn in this chapter? Deployment of wireless sensor networks is not simple and is full of unexpected problems and challenges, some of which are hard to identify and solve. Typical problems to expect are:

⇒ Node deaths and seemingly dead nodes
⇒ Interference
⇒ Missing short links, unexpected long links, or partitioned networks
⇒ Short network lifetime
⇒ Semantic (data) problems.

When testing and evaluating a new WSN application, you can use the following principles and tools:

⇒ Testing can be performed in theory, simulation, hardware testbeds, and on deployment sites.
⇒ Testing should be performed incrementally, adding complexity in each individual evaluation step.

The steps to design and implement a new WSN application are:

⇒ Analyze the requirements, including details about the environment, lifetime, the data itself, and user expectations.
⇒ Design the system top-down, by identifying first the properties of the network as a whole, followed by the node neighborhood, then individual nodes. Finally, identify the individual components of the nodes.
⇒ Derive the needed hardware and software components.
⇒ Implement and test the system bottom-up, starting from the individual hardware and software components, proceeding with the node as a whole and the network as a whole.

To conclude, you also need to keep in mind that WSN deployment is still in its baby shoes. It is a tough, complex, and badly organized process, which works well for industrial-level solutions (such as complete hardware and software solutions), but is very inefficient for new environments, applications, and research topics.

FURTHER READING

The surveys of Beutel *et al.* [1] and Oppermann *et al.* [4] offer an overview of early and recent sensor network deployments, in which typical errors and problems as well as successes are discussed. The two examples from the beginning of this chapter, the potato application and the Great Duck Island deployment, are described in the works of Langendoen *et al.* [3] and Szewczyk *et al.* [5]. The testbed INDRYIA, which has been referred several times throughout the book is described in the work of Doddavenkatappa *et al.* [2].

[1] J. Beutel, K. Römer, M. Ringwald, and M. Woehrle. *Sensor Networks: Where Theory meets Practice*, chapter Deployment Techniques for Sensor Networks. Springer, 2009.

[2] M. Doddavenkatappa, M. C. Chan, and A. L. Ananda. Indriya: A Low-Cost, 3D Wireless Sensor Network Testbed. In *Proceedings of TRIDENTCOM*, 2011.

[3] K. Langendoen, A. Baggio, and O. Visser. Murphy loves potatoes: experiences from a pilot sensor network deployment in precision agriculture. In *Proceedings of the 20th International Symposium on Parallel and Distributed Processing Symposium (IPDPS)*, page 8, Rhodes Island, Greece, 2006.

[4] F. Oppermann, C. A. Boano, and K. Römer. A Decade of Wireless Sensing Applications: Survey and Taxonomy. In Habib M. Ammari, editor, *The Art of Wireless Sensor Networks*, volume 1 of *Signals and Communication Technology*, chapter 2, pages 11–50. Springer Berlin Heidelberg, 2014.

[5] R. Szewczyk, A. Mainwaring, J. Polastre, J. Anderson, and D. Culler. An analysis of a large scale habitat monitoring application. In *Proceedings of the 2nd international conference on Embedded networked sensor systems (SenSys)*, Baltimore, MD, USA, 2004.

11

SUMMARY AND OUTLOOK

Congratulations! You can now successfully design, implement, and deploy a complete solution on your own. You are still not an expert in this topic, but you have all hands-on experience and knowledge to attack your own projects. Becoming an expert is a matter of practice and experience, which no book can provide you with. But you can start looking for more information and knowledge.

READINGS

First, you can review all of the Summary sections in this book and skim through the recommended further readings. They were kept at a minimum to actually enable you to really read or at least look at them.

STANDARDS AND SPECIFICATIONS

Next, you should become more familiar with various standards and specifications available for wireless sensor networks. This is important especially if you plan to apply sensor networks in industrial environments where similar solutions already exist to some extent.

Introduction To Wireless Sensor Networks, First Edition. Anna Förster.
© 2016 The Institute of Electrical and Electronics Engineers, Inc. Published 2016 by John Wiley & Sons, Inc.

Besides the Zigbee[1] specification, which was discussed in Chapter 3, Bluetooth (especially Bluetooth Low Energy)[2] and Wi-Fi[3] are largely used.

WirelessHART[4] and ISA100[5] share the same physical and MAC specifications (IEEE802.15.4) as Zigbee, but implement their own data dissemination protocols for different applications. The 6lowPAN[6] specification from IETF allows sending IP packets over the lower layers of 802.15.4. MiWi[7] and is yet another flavor of 802.15.4, but proprietary for Microchip microcontrollers. ANT[8] is a standard, comparable to Zigbee. It was developed by Dynastream Innovations and enables a multicast wireless sensor network for sport and health gadgets. It offers a higher transmission rate than Zigbee and has been widely used by producers such as Garmin, Adidas, and Nike for their sport tracking devices.

RFID is an important technology as well, with DASH7[9] being one of the most-used standards for RFID-enabled sensor networks.

RESEARCH

If you are interested in sensor network research and innovations, you should read the best journals and attend conferences on the subject. ACM Transactions on Sensor Networks[10] regularly publishes cutting-edge research results. Two of the best conferences for sensor networks are the ACM Conference on Embedded Sensing Systems (SenSys)[11] and the ACM/IEEE International Conference on Information Processing in Sensor Networks (IPSN).[12] The ACM International Joint Conference on Pervasive and Ubiquitous Computing (UbiComp)[13] and the IEEE International Conference on Pervasive Computing (PerCom)[14] also offer a very good overview of ongoing top-notch research.

HARDWARE

There is also the hardware market for sensor networks. Besides the Z1 and similar complete platforms, which have been used in this book, there is a very large variety

---

[1] zigbee.org
[2] www.bluetooth.com
[3] www.wi-fi.org
[4] en.hartcomm.org
[5] www.isa.org/isa100
[6] 6lowpan.net
[7] www.microchip.com/miwi
[8] www.dynastream.com/ant-wireless
[9] www.dash7-alliance.org
[10] tosn.acm.org
[11] sensys.acm.org
[12] ipsn.acm.org
[13] ubicomp.org
[14] www.percom.org

of off-the-shelf solutions for individual electronic devices, which can be integrated to form a sensor node. There is also the open hardware movement, which enables non-experts to develop their own hardware in an easy and accessible way. For example, there is Arduino,[15] which also offers many components for (wireless) sensor networks. Arduino is also suited for preparing prototypes with new hardware components and/or sensors and is highly recommended.

This book's website maintains a list of interesting hardware.

ONLINE HELP

Last but not least, this book's website offers the possibility to submit links to interesting sites, books, and articles. You can also ask questions, discuss with your peers, and leave comments and error reports about the book.

ACKNOWLEDGMENTS

Many people have helped me along the process of writing this book. Special thanks go to my beloved husband, Dr. Alexander Förster, who has motivated and supported me to continue with this project, has read most of the chapters, and made valuable suggestions and corrections.

When I started writing this book, I was still with the University of Applied Sciences in Southern Switzerland. When I finished writing this book, I was already with the University of Bremen in Germany. I am very grateful to all my colleagues from both universities: Andrea Baldassari, Jens Dede, Andreas Könsgen, Koojana Kuladinithi, Roberto Mastropietro, Kamini Garg, Silvia Giordano, Carmelita Görg, Roberto Guidi, Michela Papandrea, Daniele Puccinelli, Alessandro Puiatti, Asanga Udugama, and Karl-Heinz Volk (in alphabetical order).

However, the greatest thank goes to my students throughout all of the years when I was teaching wireless sensor networks. They all inspired me to write this book in the first place.

I hope you enjoyed reading and working with this book as much as I enjoyed writing it.

Yours,
Anna Förster

---

[15] arduino.cc

INDEX

Introduction To Wireless Sensor Networks, First Edition. Anna Förster.
© 2016 The Institute of Electrical and Electronics Engineers, Inc. Published 2016 by John Wiley & Sons, Inc.